THE MOUNTAINS OF CALIFORNIA

HOOFED LOCUSTS.

THE MOUNTAINS OF CALIFORNIA

BY

JOHN MUIR

ILLUSTRATED FROM PRELIMINARY SKETCHES AND
PHOTOGRAPHS FURNISHED BY THE AUTHOR

New and enlarged edition

TEN SPEED PRESS

TO
THE MEMORY
OF
LOUIZA STRENTZEL

CONTENTS

LIST OF ILLUSTRATIONS

LIST OF ILLUSTRATIONS

LIST OF ILLUSTRATIONS

LIST OF ILLUSTRATIONS

THE MOUNTAINS OF CALIFORNIA

THE MOUNTAINS OF CALIFORNIA

CHAPTER I

THE SIERRA NEVADA

GO where you may within the bounds of California, mountains are ever in sight, charming and glorifying every landscape. Yet so simple and massive is the topography of the State in general views, that the main central portion displays only one valley, and two chains of mountains which seem almost perfectly regular in trend and height: the Coast Range on the west side, the Sierra Nevada on the east. These two ranges coming together in curves on the north and south inclose a magnificent basin, with a level floor more than 400 miles long, and from 35 to 60 miles wide. This is the grand Central Valley of California, the waters of which have only one outlet to the sea through the Golden Gate. But with this general simplicity of features there is great complexity of hidden detail. The Coast Range, rising as a grand green barrier against the ocean, from 2000 to 8000 feet high, is composed of innumerable forest-crowned spurs, ridges, and rolling hill-waves which inclose a multitude of smaller valleys; some looking out through long, forest-lined vistas to the sea; others, with but few trees, to the Central Valley; while a thousand others yet smaller are embosomed and concealed in mild, round-browed hills, each with its own climate, soil, and productions.

Making your way through the mazes of the Coast Range to the summit of any of the inner peaks or passes opposite San Francisco, in the clear spring-time, the grandest and most telling of all California landscapes is outspread before you. At your feet lies the great Central Valley glowing golden in the sunshine, extending north and south farther than the eye can reach, one smooth, flowery, lake-like bed of fertile soil. Along its eastern margin rises the mighty Sierra, miles in height, reposing like a smooth, cumulous cloud in the sunny sky, and so gloriously colored, and so luminous, it seems to be not clothed with light, but wholly composed of it, like the wall of some celestial city. Along the top, and extending a good way down, you see a pale, pearl-gray belt of snow; and below it a belt of blue and dark purple, marking the extension of the forests; and along the base of the range a broad belt of rose-purple and yellow, where lie the miner's gold-fields and the foot-hill gardens. All these colored belts blending smoothly make a wall of light ineffably fine, and as beautiful as a rainbow, yet firm as adamant.

When I first enjoyed this superb view, one glowing April day, from the summit of the Pacheco Pass, the Central Valley, but little trampled or plowed as yet, was one furred, rich sheet of golden compositæ, and the luminous wall of the mountains shone in all its glory. Then it seemed to me the Sierra should be called not the Nevada, or Snowy Range, but the Range of Light. And after ten years spent in the heart of it, rejoicing and wondering, bathing in its glorious floods of light, seeing the sunbursts of morning among the icy peaks, the noonday radiance on the trees and rocks and snow, the flush of the alpenglow, and a thousand dashing waterfalls with their marvelous abundance of irised spray, it still seems to me above all others the Range of Light, the most divinely beautiful of all the mountain-chains I have ever seen.

The Sierra is about 500 miles long, 70 miles wide, and from 7000 to nearly 15,000 feet high. In general views no mark of man is visible on it, nor anything to suggest the richness of the life it cherishes, or the depth and grandeur of its sculpture. None of its magnificent forest-crowned ridges rises much above the general level to publish its wealth. No great valley or lake is seen, or river, or group of well-marked features of any kind, standing out in distinct pictures. Even the summit-peaks, so clear and high in the sky, seem comparatively smooth and featureless. Nevertheless, glaciers are still at work in the shadows of the peaks, and thousands of lakes and meadows shine and bloom beneath them, and the whole range is furrowed with cañons to a depth of from 2000 to 5000 feet, in which once flowed majestic glaciers, and in which now flow and sing a band of beautiful rivers.

Though of such stupendous depth, these famous cañons are not raw, gloomy, jagged-walled gorges, savage and inaccessible. With rough passages here and there they still make delightful pathways for the mountaineer, conducting from the fertile lowlands to the highest icy fountains, as a kind of mountain streets full of charming life and light, graded and sculptured by the ancient glaciers, and presenting, throughout all their courses, a rich variety of novel and attractive scenery, the most attractive that has yet been discovered in the mountain-ranges of the world.

In many places, especially in the middle region of the western flank of the range, the main cañons widen into spacious valleys or parks, diversified like artificial landscape-gardens, with charming groves and meadows, and thickets of blooming bushes, while the lofty, retiring walls, infinitely varied in form and sculpture, are fringed with ferns, flowering-plants of many species, oaks, and ever-greens, which find anchorage on a thousand narrow steps and benches; while the whole is enlivened and made glorious with rejoicing streams that come dancing and foaming over the sunny brows of the cliffs to join the shining river that flows in tranquil beauty down the middle of each one of them.

The walls of these park valleys of the Yosemite kind are made up of rocks mountains in size, partly separated from each other by narrow gorges and side-cañons; and they are so sheer in front, and so compactly built together on a level floor, that, com-prehensively seen, the parks they inclose look like immense halls or temples lighted from above. Every rock seems to glow with life. Some lean back in majestic repose; others, absolutely sheer,

MOUNT TAMALPAIS—NORTH OF THE GOLDEN GATE.

or nearly so, for thousands of feet, advance their brows in thoughtful attitudes beyond their companions, giving welcome to storms and calms alike, seemingly conscious yet heedless of everything going on about them, awful in stern majesty, types of permanence, yet associated with beauty of the frailest and most fleeting forms; their feet set in pine-groves and gay emerald meadows, their brows in the sky; bathed in light, bathed in floods of singing water, while snow-clouds, avalanches, and the winds shine and surge and wreathe about them as the years go by, as if into these mountain mansions Nature had taken pains to gather her choicest treasures to draw her lovers into close and confiding communion with her.

Here, too, in the middle region of deepest cañons are the grandest forest-trees, the Sequoia, king of conifers, the noble Sugar and Yellow Pines, Douglas Spruce, Libocedrus, and the Silver Firs, each a giant of its kind, assembled together in one and the same forest, surpassing all other coniferous forests in the world, both in the number of its species and in the size and beauty of its trees. The winds flow in melody through their colossal spires, and they are vocal everywhere with the songs of birds and running water. Miles of fragrant ceanothus and manzanita bushes bloom beneath them, and lily gardens and meadows, and damp, ferny glens in endless variety of fragrance and color, compelling the admiration of every observer. Sweeping on over ridge and valley, these noble trees extend a continuous belt from end to end of the range, only slightly interrupted by sheer-walled cañons at in-

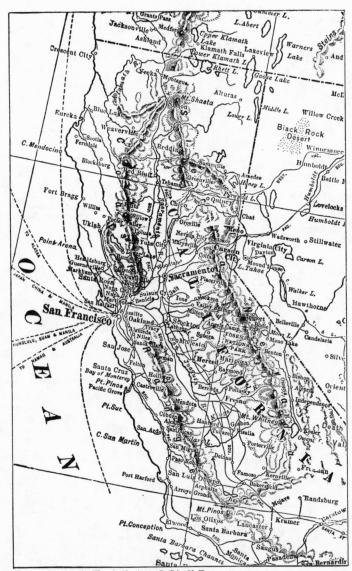

MAP OF THE SIERRA NEVADA

tervals of about fifteen and twenty miles. Here the great burly brown bears delight to roam, harmonizing with the brown boles of the trees beneath which they feed. Deer, also, dwell here, and find food and shelter in the ceanothus tangles, with a multitude of smaller people. Above this region of giants, the trees grow smaller until the utmost limit of the timber line is reached on the stormy mountain-slopes at a height of from ten to twelve thousand feet above the sea, where the Dwarf Pine is so lowly and hard beset by storms and heavy snow, it is pressed into flat tangles, over the tops of which we may easily walk. Below the main forest belt the trees likewise diminish in size, frost and burning drouth repressing and blasting alike.

The rose-purple zone along the base of the range comprehends nearly all the famous gold region of California. And here it was that miners from every country under the sun assembled in a wild, torrent-like rush to seek their fortunes. On the banks of every river, ravine, and gully they have left their marks. Every gravel- and boulder-bed has been desperately riddled over and over again. But in this region the pick and shovel, once wielded with savage enthusiasm, have been laid away, and only quartz-mining is now being carried on to any considerable extent. The zone in general is made up of low, tawny, waving foot-hills, roughened here and there with brush and trees, and outcropping masses of slate, colored gray and red with lichens. The smaller masses of slate, rising abruptly from the dry, grassy sod in leaning slabs, look like ancient tombstones in a deserted burying-ground. In early spring, say from February to April, the whole of this foot-hill belt is a paradise of bees and flowers. Refreshing rains then fall freely, birds are busy building their nests, and the sunshine is balmy and delightful. But by the end of May the soil, plants, and sky seem to have been baked in an oven. Most of the plants crumble to dust beneath the foot, and the ground is full of cracks; while the thirsty traveler gazes with eager longing through the burning glare to the snowy summits looming like hazy clouds in the distance.

The trees, mostly *Quercus Douglasii* and *Pinus Sabiniana*, thirty to forty feet high, with thin, pale-green foliage, stand far apart and cast but little shade. Lizards glide about on the rocks enjoying a constitution that no drouth can dry, and ants in amazing numbers, whose tiny sparks of life seem to burn the brighter with the increasing heat, ramble industriously in long trains in search of food. Crows, ravens, magpies—friends in distress —gather on the ground beneath the best shade-trees, panting with drooping wings and bills wide open, scarce a note from any of them during the midday hours. Quails, too, seek the shade during the heat of the day about tepid pools in the channels of the larger mid-river streams. Rabbits scurry from thicket to thicket among the ceanothus bushes, and occasionally a long-eared hare is seen cantering gracefully across the wider openings. The nights are calm and dewless during the summer, and a thousand voices proclaim the abundance of life, notwithstanding the desolating effect of dry sunshine on the plants and larger animals. The hylas make a delightfully pure and tranquil music after sunset; and coyotes, the little, despised dogs of the wilderness, brave, hardy fellows, looking like withered wisps of hay, bark in chorus for hours. Mining-towns, most of them dead, and a few living ones with bright bits of cultivation about them, occur at long intervals along the belt, and cottages covered with climbing roses, in the midst of orange and peach orchards, and sweet-scented hay-fields in fertile flats where water for irrigation may be had. But they are mostly far apart, and make scarce any mark in general views.

Every winter the High Sierra and the middle forest region get snow in glorious abundance, and even the foot-hills are at times whitened. Then all the range looks like a vast beveled wall of purest marble. The rough places are then made smooth, the death and decay of the year is covered gently and kindly, and the ground seems as clean as the sky. And though silent in its flight from the clouds, and when it is taking its place on rock, or tree, or grassy meadow, how soon the gentle snow finds a voice! Slipping from the heights, gathering in avalanches, it booms and roars like thunder, and makes a glorious show as it sweeps down the mountain-side, arrayed in long, silken streamers and wreathing, swirling films of crystal dust.

The north half of the range is mostly covered with floods of lava, and dotted with volcanoes and craters, some of them recent and perfect in form, others in various stages of decay. The south half is composed of granite nearly from base to summit, while a considerable number of peaks, in the middle of the range, are capped with metamorphic slates, among which are Mounts Dana and Gibbs to the east of Yosemite Valley. Mount Whitney, the culminating point of the range near its southern extremity, lifts its helmet-shaped crest to a height of nearly 14,700 feet. Mount Shasta, a colossal volcanic cone, rises to a height of 14,440 feet at the northern extremity, and forms a noble landmark for all the surrounding region within a radius of a hundred miles. Residual masses of volcanic rocks occur throughout most of the granitic southern portion also, and a considerable number of old volcanoes on the flanks, especially along the eastern base of the range near Mono Lake and southward. But it is only to the northward that the entire range, from base to summit, is covered with lava.

From the summit of Mount Whitney only granite is seen. Innumerable peaks and spires but little lower than its own storm-beaten crags rise in groups like forest-trees, in full view, segregated by cañons of tremendous depth and ruggedness. On Shasta nearly every feature in the vast view speaks of the old volcanic fires. Far to the northward, in Oregon, the icy volcanoes of Mount Pitt and the Three Sisters rise above the dark evergreen woods. Southward innumerable smaller craters and cones are distributed along the axis of the range and on each flank. Of these, Lassen's Butte is the highest, being nearly 11,000 feet above sea-level. Miles of its flanks are reeking and bubbling with hot springs, many of them so boisterous and sulphurous they seem ever ready to become spouting geysers like those of the Yellowstone.

The Cinder Cone near marks the most recent volcanic eruption in the Sierra. It is a symmetrical truncated cone about 700 feet high, covered with gray cinders and ashes, and has a regular unchanged crater on its summit, in which a few small Two-leaved Pines are growing. These show that the age of the cone is not less than eighty years. It stands between two lakes, which a short time ago were one. Before the cone was built, a flood of rough vesicular lava was poured into the lake, cutting it in two, and, overflowing its banks, the fiery flood advanced into the pine-woods, overwhelming the trees in its way, the charred ends of some of which may still be seen projecting from beneath the snout of the lava-stream where it came to rest. Later still there was an eruption of ashes and loose obsidian cinders, probably from the same vent, which, besides forming the Cinder Cone, scattered a heavy shower over the surrounding woods for miles to a depth of from six inches to several feet.

The history of this last Sierra eruption is also preserved in the traditions of the Pitt River Indians. They tell of a fearful time of darkness, when the sky was black with ashes and smoke that threatened every living thing with death, and that when at length the sun appeared once more it was red like blood.

Less recent craters in great numbers roughen the adjacent region; some of them with lakes in their throats, others overgrown with trees and flowers, Nature in these old hearths and firesides having literally given beauty for ashes. On the northwest side of Mount Shasta there is a subordinate cone about 3000 feet below the summit, which has been active subsequent to the breaking up of the main ice-cap that once covered the mountain, as is shown by its comparatively unwasted crater and the streams of unglaciated lava radiating from it. The main summit is about a mile and a half in diameter, bounded by small crumbling peaks and ridges, among which we seek in vain for the outlines of the ancient crater.

These ruinous masses, and the deep glacial grooves that flute the sides of the mountain, show that it has been considerably lowered and wasted by ice; how much we have no sure means of knowing. Just below the extreme summit hot sulphurous gases and vapor issue from irregular fissures, mixed with spray derived from melting snow, the last feeble expression of the mighty force that built the mountain. Not in one great convulsion was Shasta given birth. The crags of the summit and the sections exposed by the glaciers down the sides display enough of its internal framework to prove that comparatively long periods of quiescence intervened between many distinct eruptions, during which the cooling lavas ceased to flow, and became permanent additions to the bulk of the growing mountain. With alternate haste and deliberation eruption succeeded eruption till the old volcano surpassed even its present sublime height.

Standing on the icy top of this, the grandest of all the fire-mountains of the Sierra, we can hardly fail to look forward to its next eruption. Gardens, vineyards, homes have been planted confidingly on the flanks of volcanoes which, after remaining stead-

MOUNT SHASTA.

fast for ages, have suddenly blazed into violent action, and poured forth overwhelming floods of fire. It is known that more than a thousand years of cool calm have intervened between violent eruptions. Like gigantic geysers spouting molten rock instead of water, volcanoes work and rest, and we have no sure means of knowing whether they are dead when still, or only sleeping.

Along the western base of the range a telling series of sedimentary rocks containing the early history of the Sierra are now being studied. But leaving for the present these first chapters, we see that only a very short geological time ago, just before the coming on of that winter of winters called the glacial period, a vast deluge of molten rocks poured from many a chasm and crater on the flanks and summit of the range, filling lake basins and river channels, and obliterating nearly every existing feature on the northern portion. At length these all-destroying floods ceased to flow. But while the great volcanic cones built up along the axis still burned and smoked, the whole Sierra passed under the domain of ice and snow. Then over the bald, featureless, fire-blackened mountains, glaciers began to crawl, covering them from the summits to the sea with a mantle of ice; and then with infinite deliberation the work went on of sculpturing the range anew. These mighty agents of erosion, halting never through unnumbered centuries, crushed and ground the flinty lavas and granites beneath their crystal folds, wasting and building until in the fullness of time the Sierra was born again, brought to light nearly as we behold it to-day, with glaciers and snow-crushed pines at the top of the range, wheat-fields and orange-groves at the foot of it.

This change from icy darkness and death to life and beauty was slow, as we count time, and is still going on, north and south, over all the world wherever glaciers exist, whether in the form of distinct rivers, as in Switzerland, Norway, the mountains of Asia, and the Pacific Coast; or in continuous mantling folds, as in portions of Alaska, Greenland, Franz-Joseph-Land, Nova Zembla, Spitzbergen, and the lands about the South Pole. But in no country, as far as I know, may these majestic changes be studied to better advantage than in the plains and mountains of California.

Toward the close of the glacial period, when the snow-clouds became less fertile and the melting waste of sunshine became greater, the lower folds of the ice-sheet in California, discharging fleets of icebergs into the sea, began to shallow and recede from the lowlands, and then move slowly up the flanks of the Sierra in compliance with the changes of climate. The great white mantle on the mountains broke up into a series of glaciers more or less distinct and river-like, with many tributaries, and these again were melted and divided into still smaller glaciers, until now only a few of the smallest residual topmost branches of the grand system exist on the cool slopes of the summit peaks.

Plants and animals, biding their time, closely followed the retiring ice, bestowing quick and joyous animation on the new-born landscapes. Pine-trees marched up the sun-warmed moraines in long, hopeful files, taking the ground and establishing themselves as soon as it was ready for them; brown-spiked sedges fringed the shores of the new-born lakes; young rivers roared in the abandoned channels of the glaciers; flowers bloomed around the feet of the great burnished domes,—while with quick fertility mellow beds of soil, settling and warming, offered food to multitudes of Nature's waiting children, great and small, animals as well as plants; mice, squirrels, marmots, deer, bears, elephants, etc. The ground burst into bloom with magical rapidity, and the young forests into bird-song: life in every form warming and sweetening and growing richer as the years passed away over the mighty Sierra so lately suggestive of death and consummate desolation only.

It is hard without long and loving study to realize the magnitude of the work done on these mountains during the last glacial period by glaciers, which are only streams of closely compacted snow-crystals. Careful study of the phenomena presented goes to show that the pre-glacial condition of the range was comparatively simple: one vast wave of stone in which a thousand mountains, domes, cañons, ridges, etc., lay concealed. And in the development of these Nature chose for a tool not the earthquake or lightning to rend and split asunder, not the stormy torrent or eroding rain, but the tender snow-flowers noiselessly falling through unnumbered centuries, the offspring of the sun and sea. Laboring harmoniously in united strength they crushed and ground and wore away the rocks in their march, making vast beds of soil, and at the same time developed and fashioned the landscapes into the delightful variety of hill and dale and lordly mountain that mortals call beauty. Perhaps more than a mile in average depth has the range been thus degraded during the last glacial period,— a quantity of mechanical work almost inconceivably great. And our admiration must be excited again and again as we toil and study and learn that this vast job of rockwork, so far-reaching in its influences, was done by agents so fragile and small as are these flowers of the mountain clouds. Strong only by force of numbers, they carried away entire mountains, particle by particle, block by block, and cast them into the sea; sculptured, fashioned, modeled all the range, and developed its predestined beauty. All these new Sierra landscapes were evidently predestined, for the physical structure of the rocks on which the features of the scenery depend was acquired while they lay at least a mile deep below the pre-glacial surface. And it was while these features were taking form in the depths of the range, the particles of the rocks marching to their appointed places in the dark with reference to the coming beauty, that the particles of icy vapor in the sky marching to the same music assembled to bring them to the light. Then, after their grand task was done, these bands of snow-flowers, these mighty glaciers, were melted and removed as if of no more importance than dew destined to last but an hour. Few, however, of Nature's agents have left monuments so noble and enduring as they. The great granite domes a mile high, the cañons as deep, the noble peaks, the Yosemite valleys, these, and indeed

2

nearly all other features of the Sierra scenery, are glacier monuments.

Contemplating the works of these flowers of the sky, one may easily fancy them endowed with life: messengers sent down to work in the mountain mines on errands of divine love. Silently flying through the darkened air, swirling, glinting, to their appointed places, they seem to have taken counsel together, saying, " Come, we are feeble; let us help one another. We are many, and together we will be strong. Marching in close, deep ranks, let us roll away the stones from these mountain sepulchers, and set the landscapes free. Let us uncover these clustering domes. Here let us carve a lake basin; there, a Yosemite Valley; here, a channel for a river with fluted steps and brows for the plunge of song-ful cataracts. Yonder let us spread broad sheets of soil, that man and beast may be fed; and here pile trains of boulders for pines and giant Sequoias. Here make ground for a meadow; there, for a garden and grove, making it smooth and fine for small daisies and violets and beds of heathy bryanthus, spicing it well with crystals, garnet feldspar, and zircon." Thus and so on it has oftentimes seemed to me sang and planned and labored the hearty snow-flower crusaders; and nothing that I can write can possibly exaggerate the grandeur and beauty of their work. Like morning mist they have vanished in sunshine, all save the few small com-panies that still linger on the coolest mountain-sides, and, as residual glaciers, are still busily at work completing the last of the lake basins, the last beds of soil, and the sculpture of some of the highest peaks.

MOUNT HOOD.

CHAPTER II

THE GLACIERS

OF the small residual glaciers mentioned in the preceding chapter, I have found sixty-five in that portion of the range lying between latitude 36° 30′ and 39°. They occur singly or in small groups on the north sides of the peaks of the High Sierra, sheltered beneath broad frosty shadows, in amphitheaters of their own making, where the snow, shooting down from the surrounding heights in avalanches, is most abundant. Over two thirds of the entire number lie between latitude 37° and 38°, and form the highest fountains of the San Joaquin, Merced, Tuolumne, and Owen's rivers.

The glaciers of Switzerland, like those of the Sierra, are mere wasting remnants of mighty ice-floods that once filled the great valleys and poured into the sea. So, also, are those of Norway, Asia, and South America. Even the grand continuous mantles of ice that still cover Greenland, Spitzbergen, Nova Zembla, Franz-Joseph-Land, parts of Alaska, and the south polar region are shallowing and shrinking. Every glacier in the world is smaller than it once was. All the world is growing warmer, or the crop of snow-flowers is diminishing. But in contemplating the condition of the glaciers of the

20

world, we must bear in mind while trying to account for the changes going on that the same sunshine that wastes them builds them. Every glacier records the expenditure of an enormous amount of sun-heat in lifting the vapor for the snow of which it is made from the ocean to the mountains, as Tyndall strikingly shows.

The number of glaciers in the Alps, according to the Schlagintweit brothers, is 1100, of which 100 may be regarded as primary, and the total area of ice, snow, and *névé* is estimated at 1177 square miles, or an average for each glacier of little more than one square mile. On the same authority, the average height above sea-level at which they melt is about 7414 feet. The Grindelwald glacier descends below 4000 feet, and one of the Mont Blanc glaciers reaches nearly as low a point. One of the largest of the Himalaya glaciers on the head waters of the Ganges does not, according to Captain Hodgson, descend below 12,914 feet. The largest of the Sierra glaciers on Mount Shasta descends to within 9500 feet of the level of the sea, which, as far as I have observed, is the lowest point reached by any glacier within the bounds of California, the average height of all being not far from 11,000 feet.

The changes that have taken place in the glacial conditions of the Sierra from the time of greatest extension is well illustrated by the series of glaciers of every size and form extending along the mountains of the coast to Alaska. A general exploration of this instructive region shows that to the north of California, through Oregon and Washington, groups of active glaciers still exist on all the

high volcanic cones of the Cascade Range,—Mount
Pitt, the Three Sisters, Mounts Jefferson, Hood, St.
Helens, Adams, Rainier, Baker, and others,—some
of them of considerable size, though none of them
approach the sea. Of these mountains Rainier, in
Washington, is the highest and iciest. Its dome-like
summit, between 14,000 and 15,000 feet high, is
capped with ice, and eight glaciers, seven to twelve
miles long, radiate from it as a center, and form
the sources of the principal streams of the State.
The lowest-descending of this fine group flows
through beautiful forests to within 3500 feet of the
sea-level, and sends forth a river laden with glacier
mud and sand. On through British Columbia and
southeastern Alaska the broad, sustained mountain-
chain, extending along the coast, is generally glacier-
bearing. The upper branches of nearly all the main
cañons and fiords are occupied by glaciers, which
gradually increase in size, and descend lower until
the high region between Mount Fairweather and
Mount St. Elias is reached, where a considerable
number discharge into the waters of the ocean.
This is preëminently the ice-land of Alaska and of
the entire Pacific Coast.

Northward from here the glaciers gradually diminish in size and thickness, and melt at higher
levels. In Prince William Sound and Cook's Inlet
many fine glaciers are displayed, pouring from the
surrounding mountains; but to the north of latitude
62° few, if any, glaciers remain, the ground being
mostly low and the snowfall light. Between latitude 56° and 60° there are probably more than 5000
glaciers, not counting the smallest. Hundreds of

the largest size descend through the forests to the
level of the sea, or near it, though as far as my
own observations have reached, after a pretty
thorough examination of the region, not more than
twenty-five discharge icebergs into the sea. All
the long high-walled fiords into which these great
glaciers of the first class flow are of course crowded
with icebergs of every conceivable form, which are
detached with thundering noise at intervals of a
few minutes from an imposing ice-wall that is
thrust forward into deep water. But these Pacific
Coast icebergs are small as compared with those of
Greenland and the Antarctic region, and only a few
of them escape from the intricate system of channels, with which this portion of the coast is fringed,
into the open sea. Nearly all of them are swashed
and drifted by wind and tide back and forth in the
fiords until finally melted by the ocean water, the
sunshine, the warm winds, and the copious rains of
summer. Only one glacier on the coast, observed
by Prof. Russell, discharges its bergs directly into
the open sea, at Icy Cape, opposite Mount St. Elias.
The southernmost of the glaciers that reach the
sea occupies a narrow, picturesque fiord about
twenty miles to the northwest of the mouth of the
Stikeen River, in latitude 56° 50'. The fiord is called
by the natives "Hutli," or Thunder Bay, from the
noise made by the discharge of the icebergs. About
one degree farther north there are four of these
complete glaciers, discharging at the heads of the
long arms of Holkam Bay. At the head of the
Tahkoo Inlet, still farther north, there is one; and
at the head and around the sides of Glacier Bay,

trending in a general northerly direction from Cross Sound in latitude 58° to 59°, there are seven of these complete glaciers pouring bergs into the bay and its branches, and keeping up an eternal thundering. The largest of this group, the Muir, has upward of 200 tributaries, and a width below the confluence of the main tributaries of about twenty-five miles. Between the west side of this icy bay and the ocean all the ground, high and low, excepting the peaks of the Fairweather Range, is covered with a mantle of ice from 1000 to probably 3000 feet thick, which discharges by many distinct mouths.

This fragmentary ice-sheet, and the immense glaciers about Mount St. Elias, together with the multitude of separate river-like glaciers that load the slopes of the coast mountains, evidently once formed part of a continuous ice-sheet that flowed over all the region hereabouts, and only a comparatively short time ago extended as far southward as the mouth of the Strait of Juan de Fuca, probably farther. All the islands of the Alexander Archipelago, as well as the headlands and promontories of the mainland, display telling traces of this great mantle that are still fresh and unmistakable. They all have the forms of the greatest strength with reference to the action of a vast rigid press of oversweeping ice from the north and northwest, and their surfaces have a smooth, rounded, overrubbed appearance, generally free from angles. The intricate labyrinth of canals, channels, straits, passages, sounds, narrows, etc., between the islands, and extending into the mainland, of course mani-

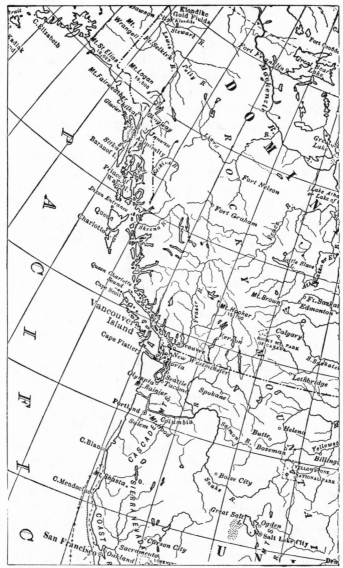

MAP OF THE GLACIER COUNTRY

fest in their forms and trends and general characteristics the same subordination to the grinding
action of universal glaciation as to their origin, and
differ from the islands and banks of the fiords only
in being portions of the pre-glacial margin of the
continent more deeply eroded, and therefore covered
by the ocean waters which flowed into them as the
ice was melted out of them. The formation and
extension of fiords in this manner is still going
on, and may be witnessed in many places in Glacier
Bay, Yakutat Bay, and adjacent regions. That the
domain of the sea is being extended over the land
by the wearing away of its shores, is well known,
but in these icy regions of Alaska, and even as far
south as Vancouver Island, the coast rocks have
been so short a time exposed to wave-action they
are but little wasted as yet. In these regions the
extension of the sea effected by its own action in
post-glacial time is scarcely appreciable as compared
with that effected by ice-action.

Traces of the vanished glaciers made during the
period of greater extension abound on the Sierra
as far south as latitude 36°. Even the polished
rock surfaces, the most evanescent of glacial records, are still found in a wonderfully perfect state
of preservation on the upper half of the middle
portion of the range, and form the most striking
of all the glacial phenomena. They occur in large
irregular patches in the summit and middle regions,
and though they have been subjected to the action
of the weather with its corroding storms for thousands of years, their mechanical excellence is such
that they still reflect the sunbeams like glass, and

attract the attention of every observer. The attention of the mountaineer is seldom arrested by moraines, however regular and high they may be, or by cañons, however deep, or by rocks, however noble in form and sculpture; but he stoops and rubs his hands admiringly on the shining surfaces and tries hard to account for their mysterious smoothness. He has seen the snow descending in avalanches, but concludes this cannot be the work of snow, for he finds it where no avalanches occur. Nor can water have done it, for he sees this smoothness glowing on the sides and tops of the highest domes. Only the winds of all the agents he knows seem capable of flowing in the directions indicated by the scoring. Indians, usually so little curious about geological phenomena, have come to me occasionally and asked me, "What makeum the ground so smooth at Lake Tenaya?" Even horses and dogs gaze wonderingly at the strange brightness of the ground, and smell the polished spaces and place their feet cautiously on them when they come to them for the first time, as if afraid of sinking. The most perfect of the polished pavements and walls lie at an elevation of from 7000 to 9000 feet above the sea, where the rock is compact silicious granite. Small dim patches may be found as low as 3000 feet on the driest and most enduring portions of sheer walls with a southern exposure, and on compact swelling bosses partially protected from rain by a covering of large boulders. On the north half of the range the striated and polished surfaces are less common, not only because this part of the chain is lower, but because the surface rocks are

chiefly porous lavas subject to comparatively rapid
waste.   The ancient moraines also, though well
preserved on most of the south half of the range,
are nearly obliterated to the northward, but their
material is found scattered and disintegrated.

A similar blurred condition of the superficial records of glacial action obtains throughout most of
Oregon, Washington, British Columbia, and Alaska,
due in great part to the action of excessive moisture.   Even in southeastern Alaska, where the most
extensive glaciers on the continent are, the more
evanescent of the traces of their former greater extension, though comparatively recent, are more obscure than those of the ancient California glaciers
where the climate is drier and the rocks more re-
sisting.

These general views of the glaciers of the Pacific
Coast will enable my readers to see something of
the changes that have taken place in California, and
will throw light on the residual glaciers of the
High Sierra.

Prior to the autumn of 1871 the glaciers of the
Sierra were unknown.   In October of that year I
discovered the Black Mountain Glacier in a
shadowy amphitheater between Black and Red
Mountains, two of the peaks of the Merced group.
This group is the highest portion of a spur that
straggles out from the main axis of the range in the
direction of Yosemite Valley.   At the time of this
interesting discovery I was exploring the *névé* amphitheaters of the group, and tracing the courses of
the ancient glaciers that once poured from its ample
fountains through the Illilouette Basin and the

Yosemite Valley, not expecting to find any active glaciers so far south in the land of sunshine.

Beginning on the northwestern extremity of the group, I explored the chief tributary basins in succession, their moraines, roches moutonnées, and splendid glacier pavements, taking them in regular succession without any reference to the time consumed in their study. The monuments of the tributary that poured its ice from between Red and Black Mountains I found to be the most interesting of them all; and when I saw its magnificent moraines extending in majestic curves from the spacious amphitheater between the mountains, I was exhilarated with the work that lay before me. It was one of the golden days of the Sierra Indian summer, when the rich sunshine glorifies every landscape however rocky and cold, and suggests anything rather than glaciers. The path of the vanished glacier was warm now, and shone in many places as if washed with silver. The tall pines growing on the moraines stood transfigured in the glowing light, the poplar groves on the levels of the basin were masses of orange-yellow, and the late-blooming goldenrods added gold to gold. Pushing on over my rosy glacial highway, I passed lake after lake set in solid basins of granite, and many a thicket and meadow watered by a stream that issues from the amphitheater and links the lakes together; now wading through plushy bogs knee-deep in yellow and purple sphagnum; now passing over bare rock. The main lateral moraines that bounded the view on either hand are from 100 to nearly 200 feet high, and about as regular as artificial embankments, and covered with a superb growth of Silver Fir and Pine. But this garden and forest luxuriance was speedily left behind. The trees were dwarfed as I ascended; patches of the alpine bryanthus and cassiope began to appear, and arctic willows pressed into flat carpets by the winter snow. The lakelets, which a few miles down the valley were so richly embroidered with flowery meadows, had here, at an elevation of 10,000 feet, only small brown mats of carex, leaving bare rocks around more than half their shores. Yet amid this alpine suppression the Mountain Pine bravely tossed his storm-beaten branches on the ledges and buttresses of Red Mountain, some specimens being over 100 feet high, and 24 feet in circumference, seemingly as fresh and vigorous as the giants of the lower zones.

Evening came on just as I got fairly within the portal of the main amphitheater. It is about a mile wide, and a little less than two miles long. The crumbling spurs and battlements of Red Mountain bound it on the north, the somber, rudely sculptured precipices of Black Mountain on the south, and a hacked, splintery *col*, curving around from mountain to mountain, shuts it in on the east.

I chose a camping-ground on the brink of one of the lakes where a thicket of Hemlock Spruce sheltered me from the night wind. Then, after making a tin-cupful of tea, I sat by my camp-fire reflecting on the grandeur and significance of the glacial records I had seen. As the night advanced the mighty rock walls of my mountain mansion seemed to come nearer, while the starry sky in glorious brightness stretched across like a ceiling from wall

MOUNT RAINIER—NORTH PUYALLUP GLACIER FROM EAGLE CLIFF.

to wall, and fitted closely down into all the spiky irregularities of the summits. Then, after a long fireside rest and a glance at my note-book, I cut a few
leafy branches for a bed, and fell into the clear,
death-like sleep of the tired mountaineer.

Early next morning I set out to trace the grand
old glacier that had done so much for the beauty
of the Yosemite region back to its farthest fountains, enjoying the charm that every explorer feels
in Nature's untrodden wildernesses. The voices of
the mountains were still asleep. The wind scarce
stirred the pine-needles. The sun was up, but it
was yet too cold for the birds and the few burrowing animals that dwell here. Only the stream, cascading from pool to pool, seemed to be wholly awake.
Yet the spirit of the opening day called to action.
The sunbeams came streaming gloriously through
the jagged openings of the *col*, glancing on the
burnished pavements and lighting the silvery lakes,
while every sun-touched rock burned white on its
edges like melting iron in a furnace. Passing round
the north shore of my camp lake I followed the central stream past many cascades from lakelet to
lakelet. The scenery became more rigidly arctic,
the Dwarf Pines and Hemlocks disappeared, and the
stream was bordered with icicles. As the sun rose
higher rocks were loosened on shattered portions of
the cliffs, and came down in rattling avalanches,
echoing wildly from crag to crag.

The main lateral moraines that extend from the
jaws of the amphitheater into the Illilouette Basin
are continued in straggling masses along the walls
of the amphitheater, while separate boulders, hun-

dreds of tons in weight, are left stranded here and there out in the middle of the channel. Here, also, I observed a series of small terminal moraines ranged along the south wall of the amphitheater, corresponding in size and form with the shadows cast by the highest portions. The meaning of this correspondence between moraines and shadows was afterward made plain. Tracing the stream back to the last of its chain of lakelets, I noticed a deposit of fine gray mud on the bottom except where the force of the entering current had prevented its settling. It looked like the mud worn from a grindstone, and I at once suspected its glacial origin, for the stream that was carrying it came gurgling out of the base of a raw moraine that seemed in process of formation. Not a plant or weather-stain was visible on its rough, unsettled surface. It is from 60 to over 100 feet high, and plunges forward at an angle of 38°. Cautiously picking my way, I gained the top of the moraine and was delighted to see a small but well characterized glacier swooping down from the gloomy precipices of Black Mountain in a finely graduated curve to the moraine on which I stood. The compact ice appeared on all the lower portions of the glacier, though gray with dirt and stones embedded in it. Farther up the ice disappeared beneath coarse granulated snow. The surface of the glacier was further characterized by dirt bands and the outcropping edges of the blue veins, showing the laminated structure of the ice. The uppermost crevasse, or " bergschrund," where the *névé* was attached to the mountain, was from 12 to 14 feet wide, and was bridged in a few places by the remains of snow avalanches. Creeping along the edge of the schrund, holding on with benumbed fingers, I discovered clear sections where the bedded structure was beautifully revealed. The surface snow, though sprinkled with stones shot down from the cliffs, was in some places almost pure, gradually becoming crystalline and changing to whitish porous ice of different shades of color, and this again changing at a depth of 20 or 30 feet to blue ice, some of the ribbon-like bands of which were nearly pure, and blended with the paler bands in the most gradual and delicate manner imaginable. A series of rugged zigzags enabled me to make my way down into the weird under-world of the crevasse. Its chambered hollows were hung with a multitude of clustered icicles, amid which pale, subdued light pulsed and shimmered with indescribable loveliness. Water dripped and tinkled overhead, and from far below came strange, solemn murmurings from currents that were feeling their way through veins and fissures in the dark. The chambers of a glacier are perfectly enchanting, notwithstanding one feels out of place in their frosty beauty. I was soon cold in my shirt-sleeves, and the leaning wall threatened to engulf me; yet it was hard to leave the delicious music of the water and the lovely light. Coming again to the surface, I noticed boulders of every size on their journeys to the terminal moraine — journeys of more than a hundred years, without a single stop, night or day, winter or summer.

The sun gave birth to a network of sweet-voiced rills that ran gracefully down the glacier, curling and swirling in their shining channels, and cut-

3

ting clear sections through the porous surface-ice into the solid blue, where the structure of the glacier was beautifully illustrated.

The series of small terminal moraines which I had observed in the morning, along the south wall of the amphitheater, correspond in every way with the moraine of this glacier, and their distribution with reference to shadows was now understood. When the climatic changes came on that caused the melting and retreat of the main glacier that filled the amphitheater, a series of residual glaciers were left in the cliff shadows, under the protection of which they lingered, until they formed the moraines we are studying. Then, as the snow became still less abundant, all of them vanished in succession, except the one just described; and the cause of its longer life is sufficiently apparent in the greater area of snow-basin it drains, and its more perfect protection from wasting sunshine. How much longer this little glacier will last depends, of course, on the amount of snow it receives from year to year, as compared with melting waste.

After this discovery, I made excursions over all the High Sierra, pushing my explorations summer after summer, and discovered that what at first sight in the distance looked like extensive snow-fields, were in great part glaciers, busily at work completing the sculpture of the summit-peaks so grandly blocked out by their giant predecessors.

On August 21, I set a series of stakes in the Maclure Glacier, near Mount Lyell, and found its rate of motion to be little more than an inch a day in the middle, showing a great contrast to the Muir

Glacier in Alaska, which, near the front, flows at a rate of from five to ten feet in twenty-four hours.

Mount Shasta has three glaciers, but Mount Whitney, although it is the highest mountain in the range, does not now cherish a single glacier. Small patches of lasting snow and ice occur on its northern slopes, but they are shallow, and present no well marked evidence of glacial motion. Its sides, however, are scored and polished in many places by the action of its ancient glaciers that flowed east and west as tributaries of the great glaciers that once filled the valleys of the Kern and Owen's rivers.

CHAPTER III

THE SNOW

THE first snow that whitens the Sierra, usually
falls about the end of October or early in November, to a depth of a few inches, after months
of the most charming Indian summer weather imaginable.   But in a few days, this light covering
mostly melts from the slopes exposed to the sun
and causes but little apprehension on the part of
mountaineers who may be lingering among the
high peaks at this time.   The first general winter
storm that yields snow that is to form a lasting
portion of the season's supply, seldom breaks on
the mountains before the end of November.   Then,
warned by the sky, cautious mountaineers, together
with the wild sheep, deer, and most of the birds
and bears, make haste to the lowlands or foot-hills;
and burrowing marmots, mountain beavers, wood-
rats, and such people go into winter quarters, some
of them not again to see the light of day until the
general awakening and resurrection of the spring in
June or July.   The first heavy fall is usually from
about two to four feet in depth.   Then, with intervals of splendid sunshine, storm succeeds storm,
heaping snow on snow, until thirty to fifty feet has
fallen.   But on account of its settling and compact-

ing, and the almost constant waste from melting
and evaporation, the average depth actually found
at any time seldom exceeds ten feet in the forest
region, or fifteen feet along the slopes of the summit peaks.

Even during the coldest weather evaporation
never wholly ceases, and the sunshine that abounds
between the storms is sufficiently powerful to melt
the surface more or less through all the winter
months. Waste from melting also goes on to some
extent on the bottom from heat stored up in the
rocks, and given off slowly to the snow in contact
with them, as is shown by the rising of the streams
on all the higher regions after the first snowfall, and
their steady sustained flow all winter.

The greater portion of the snow deposited around
the lofty summits of the range falls in small crisp
flakes and broken crystals, or, when accompanied
by strong winds and low temperature, the crystals,
instead of being locked together in their fall to
form tufted flakes, are beaten and broken into meal
and fine dust. But down in the forest region the
greater portion comes gently to the ground, light
and feathery, some of the flakes in mild weather
being nearly an inch in diameter, and it is evenly
distributed and kept from drifting to any great extent by the shelter afforded by the large trees.
Every tree during the progress of gentle storms is
loaded with fairy bloom at the coldest and darkest
time of year, bending the branches, and hushing
every singing needle. But as soon as the storm is
over, and the sun shines, the snow at once begins
to shift and settle and fall from the branches in

miniature avalanches, and the white forest soon becomes green again. The snow on the ground also
settles and thaws every bright day, and freezes at
night, until it becomes coarsely granulated, and loses
every trace of its rayed crystalline structure, and
then a man may walk firmly over its frozen surface
as if on ice. The forest region up to an elevation of
7000 feet is usually in great part free from snow in
June, but at this time the higher regions are still
heavy-laden, and are not touched by spring weather
to any considerable extent before the middle or end
of July.

One of the most striking effects of the snow on
the mountains is the burial of the rivers and small
lakes.

> As the snaw fa's in the river
> A moment white, then lost forever,

sang Burns, in illustrating the fleeting character
of human pleasure. The first snowflakes that fall
into the Sierra rivers vanish thus suddenly; but in
great storms, when the temperature is low, the
abundance of the snow at length chills the water
nearly to the freezing-point, and then, of course, it
ceases to melt and consume the snow so suddenly.
The falling flakes and crystals form cloud-like
masses of blue sludge, which are swept forward
with the current and carried down to warmer climates many miles distant, while some are lodged
against logs and rocks and projecting points of the
banks, and last for days, piled high above the level
of the water, and show white again, instead of being
at once "lost forever," while the rivers themselves

are at length lost for months during the snowy period. The snow is first built out from the banks in bossy, over-curling drifts, compacting and cementing until the streams are spanned. They then flow in the dark beneath a continuous covering across the snowy zone, which is about thirty miles wide. All the Sierra rivers and their tributaries in these high regions are thus lost every winter, as if another glacial period had come on. Not a drop of running water is to be seen excepting at a few points where large falls occur, though the rush and rumble of the heavier currents may still be heard. Toward spring, when the weather is warm during the day and frosty at night, repeated thawing and freezing and new layers of snow render the bridging-masses dense and firm, so that one may safely walk across the streams, or even lead a horse across them without danger of falling through. In June the thinnest parts of the winter ceiling, and those most exposed to sunshine, begin to give way, forming dark, rugged-edged, pit-like sinks, at the bottom of which the rushing water may be seen. At the end of June only here and there may the mountaineer find a secure snow-bridge. The most lasting of the winter bridges, thawing from below as well as from above, because of warm currents of air passing through the tunnels, are strikingly arched and sculptured; and by the occasional freezing of the oozing, dripping water of the ceiling they become brightly and picturesquely icy. In some of the reaches, where there is a free margin, we may walk through them. Small skylights appearing here and there, these tunnels are not very dark. The

roaring river fills all the arching way with impressively loud reverberating music, which is sweetened
at times by the ouzel, a bird that is not afraid to go
wherever a stream may go, and to sing wherever a
stream sings.

All the small alpine pools and lakelets are in like
manner obliterated from the winter landscapes,
either by being first frozen and then covered by
snow, or by being filled in by avalanches. The first
avalanche of the season shot into a lake basin may
perhaps find the surface frozen. Then there is a
grand crashing of breaking ice and dashing of waves
mingled with the low, deep booming of the avalanche. Detached masses of the invading snow,
mixed with fragments of ice, drift about in sludgy,
island-like heaps, while the main body of it forms
a talus with its base wholly or in part resting on
the bottom of the basin, as controlled by its depth
and the size of the avalanche. The next avalanche,
of course, encroaches still farther, and so on with
each in succession until the entire basin may be
filled and its water sponged up or displaced. This
huge mass of sludge, more or less mixed with sand,
stones, and perhaps timber, is frozen to a considerable depth, and much sun-heat is required to thaw
it. Some of these unfortunate lakelets are not
clear of ice and snow until near the end of summer.
Others are never quite free, opening only on the
side opposite the entrance of the avalanches. Some
show only a narrow crescent of water lying between
the shore and sheer bluffs of icy compacted snow,
masses of which breaking off float in front like icebergs in a miniature Arctic Ocean, while the ava-

lanche heaps leaning back against the mountains look like small glaciers. The frontal cliffs are in some instances quite picturesque, and with the berg-dotted waters in front of them lighted with sun-shine are exceedingly beautiful. It often happens that while one side of a lake basin is hopelessly snow-buried and frozen, the other, enjoying sun-shine, is adorned with beautiful flower-gardens. Some of the smaller lakes are extinguished in an in-stant by a heavy avalanche either of rocks or snow. The rolling, sliding, ponderous mass entering on one side sweeps across the bottom and up the op-posite side, displacing the water and even scraping the basin clean, and shoving the accumulated rocks and sediments up the farther bank and taking full possession. The dislodged water is in part ab-sorbed, but most of it is sent around the front of the avalanche and down the channel of the outlet, roaring and hurrying as if frightened and glad to escape.

SNOW-BANNERS

THE most magnificent storm phenomenon I ever saw, surpassing in showy grandeur the most im-posing effects of clouds, floods, or avalanches, was the peaks of the High Sierra, back of Yosemite Valley, decorated with snow-banners. Many of the starry snow-flowers, out of which these banners are made, fall before they are ripe, while most of those that do attain perfect development as six-rayed crystals glint and chafe against one another in their fall through the frosty air, and are broken into fragments. This dry fragmentary snow is still further prepared for the formation of banners by the action of the wind. For, instead of finding rest at once, like the snow which falls into the tranquil depths of the forests, it is rolled over and over, beaten against rock-ridges, and swirled in pits and hollows, like boulders, pebbles, and sand in the pot-holes of a river, until finally the delicate angles of the crystals are worn off, and the whole mass is reduced to dust. And whenever storm-winds find this prepared snow-dust in a loose condition on exposed slopes, where there is a free upward sweep to leeward, it is tossed back into the sky, and borne onward from peak to peak in the form of banners or cloudy drifts, according to the velocity of the wind and the conformation of the slopes up or around which it is driven. While thus flying through the air, a small portion makes good its escape, and remains in the sky as vapor. But far the greater part, after being driven into the sky again and again, is at length locked fast in bossy drifts, or in the wombs of glaciers, some of it to remain silent and rigid for centuries before it is finally melted and sent singing down the mountain-sides to the sea.

Yet, notwithstanding the abundance of winter snow-dust in the mountains, and the frequency of high winds, and the length of time the dust remains loose and exposed to their action, the occurrence of well-formed banners is, for causes we shall hereafter note, comparatively rare. I have seen only one display of this kind that seemed in every way perfect. This was in the winter of 1873, when the snow-laden summits were swept by a wild "norther." I happened at the time to be wintering in Yosemite Valley, that sublime Sierra temple where every day one may see the grandest sights. Yet even here the wild gala-day of the north wind seemed surpassingly glorious. I was awakened in the morning by the rocking of my cabin and the beating of pine-burs on the roof. Detached torrents and avalanches from the main wind-flood overhead were rushing wildly down the narrow side cañons, and over the precipitous walls, with loud resounding roar, rousing the pines to enthusiastic action, and making the whole valley vibrate as though it were an instrument being played.

But afar on the lofty exposed peaks of the range standing so high in the sky, the storm was expressing itself in still grander characters, which I was soon to see in all their glory. I had long been anxious to study some points in the structure of the ice-cone that is formed every winter at the foot of the upper Yosemite fall, but the blinding spray by which it is invested had hitherto prevented me from making a sufficiently near approach. This morning the entire body of the fall was torn into gauzy shreds, and blown horizontally along the face of the cliff, leaving the cone dry; and while making my way to the top of an overlooking ledge to seize so favorable an opportunity to examine the interior of the cone, the peaks of the Merced group came in sight over the shoulder of the South Dome, each waving a resplendent banner against the blue sky, as regular in form, and as firm in texture, as if woven of fine silk. So rare and splendid a phenomenon, of course, overbore all other considerations, and I at once let the ice-cone go, and began to force my way out of the valley to some dome or ridge sufficiently lofty to command a general view of the main summits, feeling assured that I should find them bannered still more gloriously; nor was I in the least disappointed. Indian Cañon, through which I climbed, was choked with snow that had been shot down in avalanches from the high cliffs on either side, rendering the ascent difficult; but inspired by the roaring storm, the tedious wallowing brought no fatigue, and in four hours I gained the top of a ridge above the valley, 8000 feet high. And there in bold relief, like a clear painting, appeared a most imposing scene. Innumerable peaks, black and sharp, rose grandly into the dark blue sky, their bases set in solid white, their sides streaked and splashed with snow, like ocean rocks with foam; and from every summit, all free and unconfused, was streaming a beautiful silky silvery banner, from half a mile to a mile in length, slender at the point of attachment, then widening gradually as it extended from the peak until it was about 1000 or 1500 feet in breadth, as near as I could estimate. The cluster of peaks called the "Crown of the Sierra," at the head of the Merced and Tuolumne rivers,—Mounts Dana, Gibbs, Conness, Lyell, Maclure, Ritter, with their nameless compeers,—each had its own refulgent banner, waving with a clearly visible motion in the sunglow, and there was not a single cloud in the sky to mar their simple grandeur. Fancy yourself standing on this Yosemite ridge looking eastward. You notice a strange

KOLANA ROCK, HETCH-HETCHY VALLEY.

garish glitter in the air. The gale drives wildly overhead with a fierce, tempestuous roar, but its violence is not felt, for you are looking through a sheltered opening in the woods as through a window. There, in the immediate foreground of your picture, rises a majestic forest of Silver Fir blooming in eternal freshness, the foliage yellow-green, and the snow beneath the trees strewn with their beautiful plumes, plucked off by the wind. Beyond, and extending over all the middle ground, are somber swaths of pine, interrupted by huge swelling ridges and domes; and just beyond the dark forest you see the monarchs of the High Sierra waving their magnificent banners. They are twenty miles away, but you would not wish them nearer, for every feature is distinct, and the whole glorious show is seen in its right proportions. After this general view, mark how sharply the dark snowless ribs and buttresses and summits of the peaks are defined, excepting the portions veiled by the banners, and how delicately their sides are streaked with snow, where it has come to rest in narrow flutings and gorges. Mark, too, how grandly the banners wave as the wind is deflected against their sides, and how trimly each is attached to the very summit of its peak, like a streamer at a masthead; how smooth and silky they are in texture, and how finely their fading fringes are penciled on the azure sky. See how dense and opaque they are at the point of attachment, and how filmy and translucent toward the end, so that the peaks back of them are seen dimly, as though you were looking through ground glass. Yet again observe how some of the longest, belonging to the loftiest summits, stream perfectly free all the way across intervening notches and passes from peak to peak, while others overlap and partly hide each other. And consider how keenly every particle of this wondrous cloth of snow is flashing out jets of light. These are the main features of the beautiful and terrible picture as seen from the forest window; and it would still be surpassingly glorious were the fore- and middle-grounds obliterated altogether, leaving only the black peaks, the white banners, and the blue sky.

Glancing now in a general way at the formation of snow-banners, we find that the main causes of the wondrous beauty and perfection of those we have been contemplating were the favorable direction and great force of the wind, the abundance of snow-dust, and the peculiar conformation of the slopes of the peaks. It is essential not only that the wind should move with great velocity and steadiness to supply a sufficiently copious and continuous stream of snow-dust, but that it should come from the north. No perfect banner is ever hung on the Sierra peaks by a south wind. Had the gale that day blown from the south, leaving other conditions unchanged, only a dull, confused, fog-like drift would have been produced; for the snow, instead of being spouted up over the tops of the peaks in concentrated currents to be drawn out as streamers, would have been shed off around the sides, and piled down into the glacier wombs. The cause of the concentrated action of the north wind is found in the peculiar form of the north sides of the peaks, where the amphitheaters of the

residual glaciers are.   In general the south sides are
convex and irregular, while the north sides are con-
cave both in their vertical and horizontal sections;
the wind in ascending these curves converges toward the summits, carrying the snow in concentrat-
ing currents with it, shooting it almost straight up
into the air above the peaks, from which it is then
carried away in a horizontal direction.

This difference in form between the north and
south sides of the peaks was almost wholly produced by the difference in the kind and quantity
of the glaciation to which they have been subjected, the north sides having been hollowed by
residual shadow-glaciers of a form that never
existed on the sun-beaten sides.

It appears, therefore, that shadows in great part
determine not only the forms of lofty icy mountains, but also those of the snow-banners that the
wild winds hang on them.

CHAPTER IV

EARLY one bright morning in the middle of Indian summer, while the glacier meadows were still crisp with frost crystals, I set out from the foot of Mount Lyell, on my way down to Yosemite Valley, to replenish my exhausted store of bread and tea. I had spent the past summer, as many preceding ones, exploring the glaciers that lie on the head waters of the San Joaquin, Tuolumne, Merced, and Owen's rivers; measuring and studying their movements, trends, crevasses, moraines, etc., and the part they had played during the period of their greater extension in the creation and development of the landscapes of this alpine wonderland. The time for this kind of work was nearly over for the year, and I began to look forward with delight to the approaching winter with its wondrous storms, when I would be warmly snow-bound in my Yosemite cabin with plenty of bread and books; but a tinge of regret came on when I considered that possibly I might not see this favorite region again until the next summer, excepting distant views from the heights about the Yosemite walls.

To artists, few portions of the High Sierra are, strictly speaking, picturesque. The whole massive uplift of the range is one great picture, not clearly divisible into smaller ones; differing much in this respect from the older, and what may be called, riper mountains of the Coast Range. All the landscapes of the Sierra, as we have seen, were born again, re-modeled from base to summit by the developing ice-floods of the last glacial winter. But all these new landscapes were not brought forth simultaneously; some of the highest, where the ice lingered longest, are tens of centuries younger than those of the warmer regions below them. In general, the younger the mountain-landscapes,—younger, I mean, with reference to the time of their emergence from the ice of the glacial period,— the less sepa-rable are they into artistic bits capable of being made into warm, sympathetic, lovable pictures with appreciable humanity in them.

Here, however, on the head waters of the Tuol-umne, is a group of wild peaks on which the geol-ogist may say that the sun has but just begun to shine, which is yet in a high degree picturesque, and in its main features so regular and evenly balanced as almost to appear conventional — one somber cluster of snow-laden peaks with gray pine-fringed granite bosses braided around its base, the whole surging free into the sky from the head of a magnificent valley, whose lofty walls are beveled away on both sides so as to embrace it all without admitting anything not strictly belonging to it. The foreground was now aflame with autumn col-ors, brown and purple and gold, ripe in the mellow sunshine; contrasting brightly with the deep, cobalt blue of the sky, and the black and gray, and pure,

4

spiritual white of the rocks and glaciers. Down through the midst, the young Tuolumne was seen pouring from its crystal fountains, now resting in glassy pools as if changing back again into ice, now leaping in white cascades as if turning to snow; gliding right and left between granite bosses, then sweeping on through the smooth, meadowy levels of the valley, swaying pensively from side to side with calm, stately gestures past dipping willows and sedges, and around groves of arrowy pine; and throughout its whole eventful course, whether flowing fast or slow, singing loud or low, ever filling the landscape with spiritual animation, and manifesting the grandeur of its sources in every movement and tone.

Pursuing my lonely way down the valley, I turned again and again to gaze on the glorious picture, throwing up my arms to inclose it as in a frame. After long ages of growth in the darkness beneath the glaciers, through sunshine and storms, it seemed now to be ready and waiting for the elected artist, like yellow wheat for the reaper; and I could not help wishing that I might carry colors and brushes with me on my travels, and learn to paint. In the mean time I had to be content with photographs on my mind and sketches in my note-books. At length, after I had rounded a precipitous headland that puts out from the west wall of the valley, every peak vanished from sight, and I pushed rapidly along the frozen meadows, over the divide between the waters of the Merced and Tuolumne, and down through the forests that clothe the slopes of Cloud's Rest, arriving in Yosemite in due time —which,

with me, is *any* time.   And, strange to say, among
the first people I met here were two artists who,
with letters of introduction, were awaiting my return.   They inquired whether in the course of my
explorations in the adjacent mountains I had ever
come upon a landscape suitable for a large painting; whereupon I began a description of the one that
had so lately excited my admiration.   Then, as I
went on further and further into details, their faces
began to glow, and I offered to guide them to it,
while they declared that they would gladly follow,
far or near, whithersoever I could spare the time to
lead them.

Since storms might come breaking down through
the fine weather at any time, burying the colors in
snow, and cutting off the artists' retreat, I advised
getting ready at once.

I led them out of the valley by the Vernal and
Nevada Falls, thence over the main dividing ridge
to the Big Tuolumne Meadows, by the old Mono
trail, and thence along the upper Tuolumne River
to its head.   This was my companions' first excursion into the High Sierra, and as I was almost always alone in my mountaineering, the way that the
fresh beauty was reflected in their faces made for
me a novel and interesting study.   They naturally
were affected most of all by the colors — the intense azure of the sky, the purplish grays of the
granite, the red and browns of dry meadows, and
the translucent purple and crimson of huckleberry
bogs; the flaming yellow of aspen groves, the silvery
flashing of the streams, and the bright green and
blue of the glacier lakes.   But the general expres-

sion of the scenery — rocky and savage — seemed
sadly disappointing; and as they threaded the forest from ridge to ridge, eagerly scanning the landscapes as they were unfolded, they said: "All this
is huge and sublime, but we see nothing as yet at
all available for effective pictures. Art is long,
and art is limited, you know; and here are foregrounds, middle-grounds, backgrounds, all alike;
bare rock-waves, woods, groves, diminutive flecks
of meadow, and strips of glittering water." " Never
mind," I replied, " only bide a wee, and I will show
you something you will like."

At length, toward the end of the second day, the
Sierra Crown began to come into view, and when we
had fairly rounded the projecting headland before
mentioned, the whole picture stood revealed in the
flush of the alpenglow. Their enthusiasm was excited beyond bounds, and the more impulsive of
the two, a young Scotchman, dashed ahead, shouting and gesticulating and tossing his arms in the
air like a madman. Here, at last, was a typical
alpine landscape.

After feasting awhile on the view, I proceeded
to make camp in a sheltered grove a little way back
from the meadow, where pine-boughs could be obtained for beds, and where there was plenty of dry
wood for fires, while the artists ran here and there,
along the river-bends and up the sides of the cañon,
choosing foregrounds for sketches. After dark,
when our tea was made and a rousing fire had been
built, we began to make our plans. They decided
to remain several days, at the least, while I con-

cluded to make an excursion in the mean time to the untouched summit of Ritter.

It was now about the middle of October, the springtime of snow-flowers. The first winter-clouds had already bloomed, and the peaks were strewn with fresh crystals, without, however, affecting the climbing to any dangerous extent. And as the weather was still profoundly calm, and the distance to the foot of the mountain only a little more than a day, I felt that I was running no great risk of being storm-bound.

Mount Ritter is king of the mountains of the middle portion of the High Sierra, as Shasta of the north and Whitney of the south sections. Moreover, as far as I know, it had never been climbed. I had explored the adjacent wilderness summer after summer, but my studies thus far had never drawn me to the top of it. Its height above sea-level is about 13,300 feet, and it is fenced round by steeply inclined glaciers, and cañons of tremendous depth and ruggedness, which render it almost inaccessible. But difficulties of this kind only exhilarate the mountaineer.

Next morning, the artists went heartily to their work and I to mine. Former experiences had given good reason to know that passionate storms, invisible as yet, might be brooding in the calm sun-gold; therefore, before bidding farewell, I warned the artists not to be alarmed should I fail to appear before a week or ten days, and advised them, in case a snow-storm should set in, to keep up big fires and shelter themselves as best they could, and

on no account to become frightened and attempt
to seek their way back to Yosemite alone through
the drifts.

My general plan was simply this: to scale the
cañon wall, cross over to the eastern flank of the
range, and then make my way southward to the
northern spurs of Mount Ritter in compliance with
the intervening topography; for to push on directly
southward from camp through the innumerable
peaks and pinnacles that adorn this portion of the
axis of the range, however interesting, would take
too much time, besides being extremely difficult and
dangerous at this time of year.

All my first day was pure pleasure; simply
mountaineering indulgence, crossing the dry pathways of the ancient glaciers, tracing happy streams,
and learning the habits of the birds and marmots
in the groves and rocks. Before I had gone a mile
from camp, I came to the foot of a white cascade
that beats its way down a rugged gorge in the
cañon wall, from a height of about nine hundred
feet, and pours its throbbing waters into the Tuolumne. I was acquainted with its fountains, which,
fortunately, lay in my course. What a fine travel-
ing companion it proved to be, what songs it sang,
and how passionately it told the mountain's own
joy! Gladly I climbed along its dashing border,
absorbing its divine music, and bathing from time
to time in waftings of irised spray. Climbing
higher, higher, new beauty came streaming on the
sight: painted meadows, late-blooming gardens,
peaks of rare architecture, lakes here and there,
shining like silver, and glimpses of the forested

middle region and the yellow lowlands far in the west. Beyond the range I saw the so-called Mono Desert, lying dreamily silent in thick purple light — a desert of heavy sun-glare beheld from a desert of ice-burnished granite. Here the waters divide, shouting in glorious enthusiasm, and falling eastward to vanish in the volcanic sands and dry sky of the Great Basin, or westward to the Great Valley of California, and thence through the Bay of San Francisco and the Golden Gate to the sea.

Passing a little way down over the summit until I had reached an elevation of about 10,000 feet, I pushed on southward toward a group of savage peaks that stand guard about Ritter on the north and west, groping my way, and dealing instinctively with every obstacle as it presented itself. Here a huge gorge would be found cutting across my path, along the dizzy edge of which I scrambled until some less precipitous point was discovered where I might safely venture to the bottom and then, selecting some feasible portion of the opposite wall, reascend with the same slow caution. Massive, flat-topped spurs alternate with the gorges, plunging abruptly from the shoulders of the snowy peaks, and planting their feet in the warm desert. These were everywhere marked and adorned with characteristic sculptures of the ancient glaciers that swept over this entire region like one vast ice-wind, and the polished surfaces produced by the ponderous flood are still so perfectly preserved that in many places the sunlight reflected from them is about as trying to the eyes as sheets of snow.

God's glacial-mills grind slowly, but they have

been kept in motion long enough in California to
grind sufficient soil for a glorious abundance of life,
though most of the grist has been carried to the
lowlands, leaving these high regions comparatively
lean and bare; while the post-glacial agents of
erosion have not yet furnished sufficient available
food over the general surface for more than a few
tufts of the hardiest plants, chiefly carices and eriogonæ.   And it is interesting to learn in this connection that the sparseness and repressed character
of the vegetation at this height is caused more by
want of soil than by harshness of climate; for, here
and there, in sheltered hollows (countersunk beneath
the general surface) into which a few rods of well-
ground moraine chips have been dumped, we find
groves of spruce and pine thirty to forty feet high,
trimmed around the edges with willow and huckleberry bushes, and oftentimes still further by an
outer ring of tall grasses, bright with lupines, larkspurs, and showy columbines, suggesting a climate
by no means repressingly severe.   All the streams,
too, and the pools at this elevation are furnished
with little gardens wherever soil can be made to lie,
which, though making scarce any show at a distance, constitute charming surprises to the appreci-
ative observer.   In these bits of leafiness a few birds
find grateful homes.   Having no acquaintance with
man, they fear no ill, and flock curiously about the
stranger, almost allowing themselves to be taken in
the hand.   In so wild and so beautiful a region was
spent my first day, every sight and sound inspiring,
leading one far out of himself, yet feeding and
building up his individuality.

Now came the solemn, silent evening. Long, blue, spiky shadows crept out across the snow-fields, while a rosy glow, at first scarce discernible, gradually deepened and suffused every mountain-top, flushing the glaciers and the harsh crags above them. This was the alpenglow, to me one of the most impressive of all the terrestrial manifestations of God. At the touch of this divine light, the mountains seemed to kindle to a rapt, religious consciousness, and stood hushed and waiting like devout worshipers. Just before the alpenglow began to fade, two crimson clouds came streaming across the summit like wings of flame, rendering the sublime scene yet more impressive; then came darkness and the stars.

Icy Ritter was still miles away, but I could proceed no farther that night. I found a good campground on the rim of a glacier basin about 11,000 feet above the sea. A small lake nestles in the bottom of it, from which I got water for my tea, and a stormbeaten thicket near by furnished abundance of resiny fire-wood. Somber peaks, hacked and shattered, circled half-way around the horizon, wearing a savage aspect in the gloaming, and a waterfall chanted solemnly across the lake on its way down from the foot of a glacier. The fall and the lake and the glacier were almost equally bare; while the scraggy pines anchored in the rock-fissures were so dwarfed and shorn by storm-winds that you might walk over their tops. In tone and aspect the scene was one of the most desolate I ever beheld. But the darkest scriptures of the mountains are illumined with bright passages of love that never fail to make themselves felt when one is alone.

I made my bed in a nook of the pine-thicket, where the branches were pressed and crinkled overhead like a roof, and bent down around the sides. These are the best bedchambers the high mountains afford—snug as squirrel-nests, well ventilated, full of spicy odors, and with plenty of wind-played needles to sing one asleep. I little expected company, but, creeping in through a low side-door, I found five or six birds nestling among the tassels. The night-wind began to blow soon after dark; at first only a gentle breathing, but increasing toward midnight to a rough gale that fell upon my leafy roof in ragged surges like a cascade, bearing wild sounds from the crags overhead. The waterfall sang in chorus, filling the old ice-fountain with its solemn roar, and seeming to increase in power as the night advanced—fit voice for such a landscape. I had to creep out many times to the fire during the night, for it was biting cold and I had no blankets. Gladly I welcomed the morning star.

The dawn in the dry, wavering air of the desert was glorious. Everything encouraged my undertaking and betokened success. There was no cloud in the sky, no storm-tone in the wind. Breakfast of bread and tea was soon made. I fastened a hard, durable crust to my belt by way of provision, in case I should be compelled to pass a night on the mountain-top; then, securing the remainder of my little stock against wolves and wood-rats, I set forth free and hopeful.

How glorious a greeting the sun gives the mountains! To behold this alone is worth the pains of any excursion a thousand times over. The highest peaks burned like islands in a sea of liquid shade. Then the lower peaks and spires caught the glow, and long lances of light, streaming through many a notch and pass, fell thick on the frozen meadows. The majestic form of Ritter was full in sight, and I pushed rapidly on over rounded rock-bosses and pavements, my iron-shod shoes making a clanking sound, suddenly hushed now and then in rugs of bryanthus, and sedgy lake-margins soft as moss. Here, too, in this so-called "land of desolation," I met cassiope, growing in fringes among the battered rocks. Her blossoms had faded long ago, but they were still clinging with happy memories to the evergreen sprays, and still so beautiful as to thrill every fiber of one's being. Winter and summer, you may hear her voice, the low, sweet melody of her purple bells. No evangel among all the mountain plants speaks Nature's love more plainly than cassiope. Where she dwells, the redemption of the coldest solitude is complete. The very rocks and glaciers seem to feel her presence, and become imbued with her own fountain sweetness. All things were warming and awakening. Frozen rills began to flow, the marmots came out of their nests in boulder-piles and climbed sunny rocks to bask, and the dun-headed sparrows were flitting about seeking their breakfasts. The lakes seen from every ridge-top were brilliantly rippled and spangled, shimmering like the thickets of the low Dwarf Pines. The rocks, too, seemed responsive to the vital heat—rock-crystals and snow-crystals thrill-

ing alike.   I strode on exhilarated, as if never more
to feel fatigue, limbs moving of themselves, every
sense unfolding like the thawing flowers, to take
part in the new day harmony.

All along my course thus far, excepting when
down in the cañons, the landscapes were mostly
open to me, and expansive, at least on one side.
On the left were the purple plains of Mono, reposing dreamily and warm; on the right, the near
peaks springing keenly into the thin sky with
more and more impressive sublimity.   But these
larger views were at length lost. Rugged spurs,
and moraines, and huge, projecting buttresses
began to shut me in.   Every feature became more
rigidly alpine, without, however, producing any
chilling effect; for going to the mountains is like
going home.   We always find that the strangest
objects in these fountain wilds are in some degree
familiar, and we look upon them with a vague
sense of having seen them before.

On the southern shore of a frozen lake, I encountered an extensive field of hard, granular
snow, up which I scampered in fine tone, intending to follow it to its head, and cross the rocky
spur against which it leans, hoping thus to come
direct upon the base of the main Ritter peak. The
surface was pitted with oval hollows, made by
stones and drifted pine-needles that had melted
themselves into the mass by the radiation of absorbed sun-heat.   These afforded good footholds,
but the surface curved more and more steeply at
the head, and the pits became shallower and less
abundant, until I found myself in danger of being

GENERAL GRANT TREE—GENERAL GRANT NATIONAL PARK.

shed off like avalanching snow. I persisted, however, creeping on all fours, and shuffling up the
smoothest places on my back, as I had often done
on burnished granite, until, after slipping several
times, I was compelled to retrace my course to
the bottom, and make my way around the west
end of the lake, and thence up to the summit of
the divide between the head waters of Rush Creek
and the northernmost tributaries of the San
Joaquin.

Arriving on the summit of this dividing crest,
one of the most exciting pieces of pure wilderness
was disclosed that I ever discovered in all my
mountaineering. There, immediately in front,
loomed the majestic mass of Mount Ritter, with a
glacier swooping down its face nearly to my feet,
then curving westward and pouring its frozen flood
into a dark blue lake, whose shores were bound
with precipices of crystalline snow; while a deep
chasm drawn between the divide and the glacier
separated the massive picture from everything else.
I could see only the one sublime mountain, the one
glacier, the one lake; the whole veiled with one
blue shadow—rock, ice, and water close together
without a single leaf or sign of life. After gazing
spellbound, I began instinctively to scrutinize every
notch and gorge and weathered buttress of the
mountain, with reference to making the ascent.
The entire front above the glacier appeared as one
tremendous precipice, slightly receding at the top,
and bristling with spires and pinnacles set above
one another in formidable array. Massive lichen-
stained battlements stood forward here and there,

hacked at the top with angular notches, and separated by frosty gullies and recesses that have been
veiled in shadow ever since their creation; while
to right and left, as far as I could see, were huge,
crumbling buttresses, offering no hope to the
climber. The head of the glacier sends up a few
finger-like branches through narrow *couloirs;* but
these seemed too steep and short to be available,
especially as I had no ax with which to cut steps,
and the numerous narrow-throated gullies down
which stones and snow are avalanched seemed hopelessly steep, besides being interrupted by vertical
cliffs; while the whole front was rendered still more
terribly forbidding by the chill shadow and the
gloomy blackness of the rocks.

Descending the divide in a hesitating mood, I
picked my way across the yawning chasm at the
foot, and climbed out upon the glacier. There
were no meadows now to cheer with their brave
colors, nor could I hear the dun-headed sparrows,
whose cheery notes so often relieve the silence of
our highest mountains. The only sounds were the
gurgling of small rills down in the veins and crevasses of the glacier, and now and then the rattling
report of falling stones, with the echoes they shot out
into the crisp air.

I could not distinctly hope to reach the summit
from this side, yet I moved on across the glacier as
if driven by fate. Contending with myself, the
season is too far spent, I said, and even should I be
successful, I might be storm-bound on the mountain; and in the cloud-darkness, with the cliffs and
crevasses covered with snow, how could I escape?

No; I must wait till next summer. I would only approach the mountain now, and inspect it, creep about its flanks, learn what I could of its history, holding myself ready to flee on the approach of the first storm-cloud. But we little know until tried how much of the uncontrollable there is in us, urging across glaciers and torrents, and up dangerous heights, let the judgment forbid as it may.

I succeeded in gaining the foot of the cliff on the eastern extremity of the glacier, and there discovered the mouth of a narrow avalanche gully, through which I began to climb, intending to follow it as far as possible, and at least obtain some fine wild views for my pains. Its general course is oblique to the plane of the mountain-face, and the metamorphic slates of which the mountain is built are cut by cleavage planes in such a way that they weather off in angular blocks, giving rise to irregular steps that greatly facilitate climbing on the sheer places. I thus made my way into a wilderness of crumbling spires and battlements, built together in bewildering combinations, and glazed in many places with a thin coating of ice, which I had to hammer off with stones. The situation was becoming gradually more perilous; but, having passed several dangerous spots, I dared not think of descending; for, so steep was the entire ascent, one would inevitably fall to the glacier in case a single misstep were made. Knowing, therefore, the tried danger beneath, I became all the more anxious concerning the developments to be made above, and began to be conscious of a vague foreboding of what actually befell; not that I was given to fear, but rather because my instincts, usually so positive and true, seemed vitiated in some way, and were leading me astray. At length, after attaining an elevation of about 12,800 feet, I found myself at the foot of a sheer drop in the bed of the avalanche channel I was tracing, which seemed absolutely to bar further progress. It was only about forty-five or fifty feet high, and somewhat roughened by fissures and projections; but these seemed so slight and insecure, as footholds, that I tried hard to avoid the precipice altogether, by scaling the wall of the channel on either side. But, though less steep, the walls were smoother than the obstructing rock, and repeated efforts only showed that I must either go right ahead or turn back. The tried dangers beneath seemed even greater than that of the cliff in front; therefore, after scanning its face again and again, I began to scale it, picking my holds with intense caution. After gaining a point about halfway to the top, I was suddenly brought to a dead stop, with arms outspread, clinging close to the face of the rock, unable to move hand or foot either up or down. My doom appeared fixed. I *must* fall. There would be a moment of bewilderment, and then a lifeless rumble down the one general precipice to the glacier below.

When this final danger flashed upon me, I became nerve-shaken for the first time since setting foot on the mountains, and my mind seemed to fill with a stifling smoke. But this terrible eclipse lasted only a moment, when life blazed forth again with preternatural clearness. I seemed suddenly to become possessed of a new sense. The other self, bygone experiences, Instinct, or Guardian Angel,— call it what you will,— came forward and assumed control. Then my trembling muscles became firm again, every rift and flaw in the rock was seen as through a microscope, and my limbs moved with a positiveness and precision with which I seemed to have nothing at all to do. Had I been borne aloft upon wings, my deliverance could not have been more complete.

Above this memorable spot, the face of the mountain is still more savagely hacked and torn. It is a maze of yawning chasms and gullies, in the angles of which rise beetling crags and piles of detached boulders that seem to have been gotten ready to be launched below. But the strange influx of strength I had received seemed inexhaustible. I found a way without effort, and soon stood upon the topmost crag in the blessed light.

How truly glorious the landscape circled around this noble summit!—giant mountains, valleys innumerable, glaciers and meadows, rivers and lakes, with the wide blue sky bent tenderly over them all. But in my first hour of freedom from that terrible shadow, the sunlight in which I was laving seemed all in all.

Looking southward along the axis of the range, the eye is first caught by a row of exceedingly sharp and slender spires, which rise openly to a height of about a thousand feet, above a series of short, residual glaciers that lean back against their bases; their fantastic sculpture and the unrelieved sharpness with which they spring out of the ice rendering them peculiarly wild and striking. These

5

are " The Minarets." Beyond them you behold a
sublime wilderness of mountains, their snowy summits towering together in crowded abundance, peak
beyond peak, swelling higher, higher as they sweep
on southward, until the culminating point of the
range is reached on Mount Whitney, near the head
of the Kern River, at an elevation of nearly 14,700
feet above the level of the sea.

Westward, the general flank of the range is seen
flowing sublimely away from the sharp summits,
in smooth undulations; a sea of huge gray granite
waves dotted with lakes and meadows, and fluted
with stupendous cañons that grow steadily deeper
as they recede in the distance. Below this gray
region lies the dark forest zone, broken here and
there by upswelling ridges and domes; and yet
beyond lies a yellow, hazy belt, marking the broad
plain of the San Joaquin, bounded on its farther
side by the blue mountains of the coast.

Turning now to the northward, there in the immediate foreground is the glorious Sierra Crown,
with Cathedral Peak, a temple of marvelous architecture, a few degrees to the left of it; the gray,
massive form of Mammoth Mountain to the right;
while Mounts Ord, Gibbs, Dana, Conness, Tower
Peak, Castle Peak, Silver Mountain, and a host of
noble companions, as yet nameless, make a sublime show along the axis of the range.

Eastward, the whole region seems a land of desolation covered with beautiful light. The torrid
volcanic basin of Mono, with its one bare lake
fourteen miles long; Owen's Valley and the broad
lava table-land at its head, dotted with craters, and

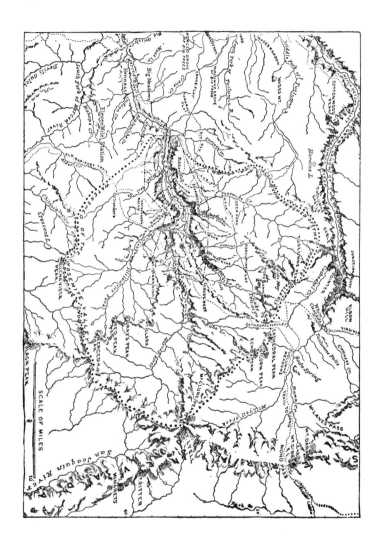

the massive Inyo Range, rivaling even the Sierra in height; these are spread, map-like, beneath you, with countless ranges beyond, passing and overlapping one another and fading on the glowing horizon.

At a distance of less than 3,000 feet below the summit of Mount Ritter you may find tributaries of the San Joaquin and Owen's rivers, bursting forth from the ice and snow of the glaciers that load its flanks; while a little to the north of here are found the highest affluents of the Tuolumne and Merced. Thus, the fountains of four of the principal rivers of California are within a radius of four or five miles.

Lakes are seen gleaming in all sorts of places,— round, or oval, or square, like very mirrors; others narrow and sinuous, drawn close around the peaks like silver zones, the highest reflecting only rocks, snow, and the sky. But neither these nor the glaciers, nor the bits of brown meadow and moorland that occur here and there, are large enough to make any marked impression upon the mighty wilderness of mountains. The eye, rejoicing in its freedom, roves about the vast expanse, yet returns again and again to the fountain peaks. Perhaps some one of the multitude excites special attention, some gigantic castle with turret and battlement, or some Gothic cathedral more abundantly spired than Milan's. But, generally, when looking for the first time from an all-embracing standpoint like this, the inexperienced observer is oppressed by the incomprehensible grandeur, variety, and abundance of the mountains rising shoulder to shoulder beyond the reach of vision; and it is only after they have been studied one by one, long and lovingly, that their far-reaching harmonies become manifest. Then, penetrate the wilderness where you may, the main telling features, to which all the surrounding topography is subordinate, are quickly perceived, and the most complicated clusters of peaks stand revealed harmoniously correlated and fashioned like works of art — eloquent monuments of the ancient ice-rivers that brought them into relief from the general mass of the range. The cañons, too, some of them a mile deep, mazing wildly through the mighty host of mountains, however lawless and ungovernable at first sight they appear, are at length recognized as the necessary effects of causes which followed each other in harmonious sequence — Nature's poems carved on tables of stone — the simplest and most emphatic of her glacial compositions.

Could we have been here to observe during the glacial period, we should have overlooked a wrinkled ocean of ice as continuous as that now covering the landscapes of Greenland; filling every valley and cañon with only the tops of the fountain peaks rising darkly above the rock-encumbered ice-waves like islets in a stormy sea—those islets the only hints of the glorious landscapes now smiling in the sun. Standing here in the deep, brooding silence all the wilderness seems motionless, as if the work of creation were done. But in the midst of this outer steadfastness we know there is incessant motion and change. Ever and anon, avalanches are falling from yonder peaks. These cliff-bound glaciers, seemingly wedged and immovable, are flowing like water and grinding the rocks beneath them. The lakes are lapping their granite shores and wearing them away, and every one of these rills and young rivers is fretting the air into music, and carrying the mountains to the plains. Here are the roots of all the life of the valleys, and here more simply than elsewhere is the eternal flux of nature manifested. Ice changing to water, lakes to meadows, and mountains to plains. And while we thus contemplate Nature's methods of landscape creation, and, reading the records she has carved on the rocks, reconstruct, however imperfectly, the landscapes of the past, we also learn that as these we now behold have succeeded those of the preglacial age, so they in turn are withering and vanishing to be succeeded by others yet unborn.

But in the midst of these fine lessons and landscapes, I had to remember that the sun was wheeling far to the west, while a new way down the mountain had to be discovered to some point on the timber line where I could have a fire; for I had not even burdened myself with a coat. I first scanned the western spurs, hoping some way might appear through which I might reach the northern glacier, and cross its snout; or pass around the lake into which it flows, and thus strike my morning track. This route was soon sufficiently unfolded to show that, if practicable at all, it would require so much time that reaching camp that night would be out of the question. I therefore scrambled back eastward, descending the southern slopes obliquely at the same time. Here the crags seemed less formid-

able, and the head of a glacier that flows northeast came in sight, which I determined to follow as
far as possible, hoping thus to make my way to
the foot of the peak on the east side, and thence
across the intervening cañons and ridges to camp.

The inclination of the glacier is quite moderate
at the head, and, as the sun had softened the *névé*,
I made safe and rapid progress, running and sliding,
and keeping up a sharp outlook for crevasses.
About half a mile from the head, there is an ice-
cascade, where the glacier pours over a sharp declivity and is shattered into massive blocks separated by deep, blue fissures. To thread my way
through the slippery mazes of this crevassed portion seemed impossible, and I endeavored to avoid
it by climbing off to the shoulder of the mountain.
But the slopes rapidly steepened and at length fell
away in sheer precipices, compelling a return to the
ice. Fortunately, the day had been warm enough
to loosen the ice-crystals so as to admit of hollows
being dug in the rotten portions of the blocks, thus
enabling me to pick my way with far less difficulty
than I had anticipated. Continuing down over the
snout, and along the left lateral moraine, was only
a confident saunter, showing that the ascent of the
mountain by way of this glacier is easy, provided
one is armed with an ax to cut steps here and there.

The lower end of the glacier was beautifully
waved and barred by the outcropping edges of the
bedded ice-layers which represent the annual snowfalls, and to some extent the irregularities of structure caused by the weathering of the walls of crevasses, and by separate snowfalls which have been

followed by rain, hail, thawing and freezing, etc. Small rills were gliding and swirling over the melting surface with a smooth, oily appearance, in channels of pure ice—their quick, compliant movements contrasting most impressively with the rigid, invisible flow of the glacier itself, on whose back they all were riding.

Night drew near before I reached the eastern base of the mountain, and my camp lay many a rugged mile to the north; but ultimate success was assured. It was now only a matter of endurance and ordinary mountain-craft. The sunset was, if possible, yet more beautiful than that of the day before. The Mono landscape seemed to be fairly saturated with warm, purple light. The peaks marshaled along the summit were in shadow, but through every notch and pass streamed vivid sun-fire, soothing and irradiating their rough, black angles, while companies of small, luminous clouds hovered above them like very angels of light.

Darkness came on, but I found my way by the trends of the cañons and the peaks projected against the sky. All excitement died with the light, and then I was weary. But the joyful sound of the waterfall across the lake was heard at last, and soon the stars were seen reflected in the lake itself. Taking my bearings from these, I discovered the little pine thicket in which my nest was, and then I had a rest such as only a tired mountaineer may enjoy. After lying loose and lost for awhile, I made a sunrise fire, went down to the lake, dashed water on my head, and dipped a cupful for tea. The revival brought about by bread and tea was as complete as the exhaustion

from excessive enjoyment and toil. Then I crept
beneath the pine-tassels to bed. The wind was
frosty and the fire burned low, but my sleep was
none the less sound, and the evening constellations
had swept far to the west before I awoke.

After thawing and resting in the morning sunshine, I sauntered home,—that is, back to the Tuolumne camp,—bearing away toward a cluster of
peaks that hold the fountain snows of one of the
north tributaries of Rush Creek. Here I discovered
a group of beautiful glacier lakes, nestled together in a grand amphitheater. Toward evening, I
crossed the divide separating the Mono waters
from those of the Tuolumne, and entered the
glacier basin that now holds the fountain snows
of the stream that forms the upper Tuolumne cascades. This stream I traced down through its
many dells and gorges, meadows and bogs, reaching the brink of the main Tuolumne at dusk.

A loud whoop for the artists was answered again
and again. Their camp-fire came in sight, and
half an hour afterward I was with them. They
seemed unreasonably glad to see me. I had been
absent only three days; nevertheless, though the
weather was fine, they had already been weighing
chances as to whether I would ever return, and
trying to decide whether they should wait longer
or begin to seek their way back to the lowlands.
Now their curious troubles were over. They
packed their precious sketches, and next morning
we set out homeward bound, and in two days
entered the Yosemite Valley from the north by
way of Indian Cañon.

CHAPTER V

THE sustained grandeur of the High Sierra is strikingly illustrated by the great height of the passes. Between latitude 36° 20′ and 38° the lowest pass, gap, gorge, or notch of any kind cutting across the axis of the range, as far as I have discovered, exceeds 9000 feet in height above the level of the sea; while the average height of all that are in use, either by Indians or whites, is perhaps not less than 11,000 feet, and not one of these is a carriage-pass.

Farther north a carriage-road has been constructed through what is known as the Sonora Pass, on the head waters of the Stanislaus and Walker's rivers, the summit of which is about 10,000 feet above the sea. Substantial wagon-roads have also been built through the Carson and Johnson passes, near the head of Lake Tahoe, over which immense quantities of freight were hauled from California to the mining regions of Nevada, before the construction of the Central Pacific Railroad.

Still farther north a considerable number of comparatively low passes occur, some of which are accessible to wheeled vehicles, and through these rugged defiles during the exciting years of the gold

74

period long emigrant-trains with foot-sore cattle wearily toiled. After the toil-worn adventurers had escaped a thousand dangers and had crawled thousands of miles across the plains the snowy Sierra at last loomed in sight, the eastern wall of the land of gold. And as with shaded eyes they gazed through the tremulous haze of the desert, with what joy must they have descried the pass through which they were to enter the better land of their hopes and dreams!

Between the Sonora Pass and the southern extremity of the High Sierra, a distance of nearly 160 miles, there are only five passes through which trails conduct from one side of the range to the other. These are barely practicable for animals; a pass in these regions meaning simply any notch or cañon through which one may, by the exercise of unlimited patience, make out to lead a mule, or a sure-footed mustang; animals that can slide or jump as well as walk. Only three of the five passes may be said to be in use, viz.: the Kearsarge, Mono, and Virginia Creek; the tracks leading through the others being only obscure Indian trails, not graded in the least, and scarcely traceable by white men; for much of the way is over solid rock and earthquake avalanche taluses, where the unshod ponies of the Indians leave no appreciable sign. Only skilled mountaineers are able to detect the marks that serve to guide the Indians, such as slight abrasions of the looser rocks, the displacement of stones here and there, and bent bushes and weeds. A general knowledge of the topography is, then, the main guide, enabling one to determine where

the trail ought to go—*must* go. One of these Indian trails crosses the range by a nameless pass
between the head waters of the south and middle
forks of the San Joaquin, the other between the
north and middle forks of the same river, just to
the south of "The Minarets"; this last being about
9000 feet high, is the lowest of the five. The Kearsarge is the highest, crossing the summit near the
head of the south fork of King's River, about eight
miles to the north of Mount Tyndall, through the
midst of the most stupendous rock-scenery. The
summit of this pass is over 12,000 feet above sea-
level; nevertheless, it is one of the safest of the five,
and is used every summer, from July to October
or November, by hunters, prospectors, and stock-
owners, and to some extent by enterprising pleasure-
seekers also. For, besides the surpassing grandeur
of the scenery about the summit, the trail, in ascending the western flank of the range, conducts
through a grove of the giant Sequoias, and through
the magnificent Yosemite Valley of the south fork
of King's River. This is, perhaps, the highest traveled pass on the North American continent.

The Mono Pass lies to the east of Yosemite Valley, at the head of one of the tributaries of the
south fork of the Tuolumne. This is the best
known and most extensively traveled of all that
exist in the High Sierra. A trail was made through
it about the time of the Mono gold excitement, in
the year 1858, by adventurous miners and prospectors—men who would build a trail down the throat
of darkest Erebus on the way to gold. Though
more than a thousand feet lower than the Kear-

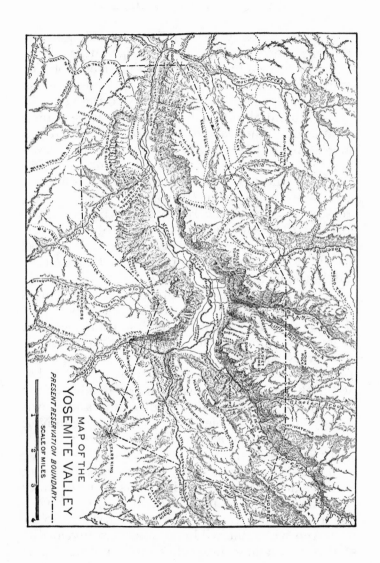

MAP OF THE
YOSEMITE VALLEY

PRESENT RESERVATION BOUNDARY.———

SCALE OF MILES

sarge, it is scarcely less sublime in rock-scenery, while in snowy, falling water it far surpasses it. Being so favorably situated for the stream of Yosemite travel, the more adventurous tourists cross over through this glorious gateway to the volcanic region around Mono Lake. It has therefore gained a name and fame above every other pass in the range. According to the few barometrical observations made upon it, its highest point is 10,765 feet above the sea. The other pass of the five we have been considering is somewhat lower, and crosses the axis of the range a few miles to the north of the Mono Pass, at the head of the southernmost tributary of Walker's River. It is used chiefly by roaming bands of the Pah Ute Indians and "sheepmen."

But, leaving wheels and animals out of the question, the free mountaineer with a sack of bread on his shoulders and an ax to cut steps in ice and frozen snow can make his way across the range almost everywhere, and at any time of year when the weather is calm. To him nearly every notch between the peaks is a pass, though much patient step-cutting is at times required up and down steeply inclined glaciers, with cautious climbing over precipices that at first sight would seem hopelessly inaccessible.

In pursuing my studies, I have crossed from side to side of the range at intervals of a few miles all along the highest portion of the chain, with far less real danger than one would naturally count on. And what fine wildness was thus revealed— storms and avalanches, lakes and waterfalls, gardens and meadows, and interesting animals— only those will ever know who give the freest and most buoyant portion of their lives to climbing and seeing for themselves.

To the timid traveler, fresh from the sedimentary levels of the lowlands, these highways, however picturesque and grand, seem terribly forbidding—cold, dead, gloomy gashes in the bones of the mountains, and of all Nature's ways the ones to be most cautiously avoided. Yet they are full of the finest and most telling examples of Nature's love; and though hard to travel, none are safer. For they lead through regions that lie far above the ordinary haunts of the devil, and of the pestilence that walks in darkness. True, there are innumerable places where the careless step will be the last step; and a rock falling from the cliffs may crush without warning like lightning from the sky; but what then? Accidents in the mountains are less common than in the lowlands, and these mountain mansions are decent, delightful, even divine, places to die in, compared with the doleful chambers of civilization. Few places in this world are more dangerous than home. Fear not, therefore, to try the mountain-passes. They will kill care, save you from deadly apathy, set you free, and call forth every faculty into vigorous, enthusiastic action. Even the sick should try these so-called dangerous passes, because for every unfortunate they kill, they cure a thousand.

All the passes make their steepest ascents on the eastern flank. On this side the average rise is not far from a thousand feet to the mile, while on the west it is about two hundred feet. Another marked difference between the eastern and western portions of the passes is that the former begin at the very foot of the range, while the latter can hardly be said to begin lower than an elevation of from seven to ten thousand feet. Approaching the range from the gray levels of Mono and Owen's Valley on the east, the traveler sees before him the steep, short passes in full view, fenced in by rugged spurs that come plunging down from the shoulders of the peaks on either side, the courses of the more direct being disclosed from top to bottom without interruption. But from the west one sees nothing of the way he may be seeking until near the summit, after days have been spent in threading the forests growing on the main dividing ridges between the river cañons.

It is interesting to observe how surely the alp-crossing animals of every kind fall into the same trails. The more rugged and inaccessible the general character of the topography of any particular region, the more surely will the trails of white men, Indians, bears, wild sheep, etc., be found converging into the best passes. The Indians of the western slope venture cautiously over the passes in settled weather to attend dances, and obtain loads of pine-nuts and the larvæ of a small fly that breeds in Mono and Owen's lakes, which, when dried, forms an important article of food; while the Pah Utes cross over from the east to hunt the deer and obtain supplies of acorns; and it is truly astonishing to see what immense loads the haggard old squaws make out to carry barefooted through these rough passes, oftentimes for a distance of sixty or seventy miles. They are always accompanied by the men, who stride on, unburdened and erect, a little in advance, kindly stooping at difficult places to pile stepping-stones for their patient, pack-animal wives, just as they would prepare the way for their ponies.

Bears evince great sagacity as mountaineers, but although they are tireless and enterprising travelers they seldom cross the range. I have several times tracked them through the Mono Pass, but only in late years, after cattle and sheep had passed that way, when they doubtless were following to feed on the stragglers and on those that had been killed by falling over the rocks. Even the wild sheep, the best mountaineers of all, choose regular passes in making journeys across the summits. Deer seldom cross the range in either direction. I have never yet observed a single specimen of the mule-deer of the Great Basin west of the summit, and rarely one of the black-tailed species on the eastern slope, notwithstanding many of the latter ascend the range nearly to the summit every summer, to feed in the wild gardens and bring forth their young.

The glaciers are the pass-makers, and it is by them that the courses of all mountaineers are predestined. Without exception every pass in the Sierra was created by them without the slightest aid or predetermining guidance from any of the cataclysmic agents. I have seen elaborate statements of the amount of drilling and blasting accomplished in the construction of the railroad

6

across the Sierra, above Donner Lake; but for every pound of rock moved in this way, the glaciers which descended east and west through this same pass, crushed and carried away more than a hundred tons.

The so-called practicable road-passes are simply those portions of the range more degraded by glacial action than the adjacent portions, and degraded in such a way as to leave the summits rounded, instead of sharp; while the peaks, from the superior strength and hardness of their rocks, or from more favorable position, having suffered less degradation, are left towering above the passes as if they had been heaved into the sky by some force acting from beneath.

The scenery of all the passes, especially at the head, is of the wildest and grandest description,— lofty peaks massed together and laden around their bases with ice and snow; chains of glacier lakes; cascading streams in endless variety, with glorious views, westward over a sea of rocks and woods, and eastward over strange ashy plains, volcanoes, and the dry, dead-looking ranges of the Great Basin. Every pass, however, possesses treasures of beauty all its own.

Having thus in a general way indicated the height, leading features, and distribution of the principal passes, I will now endeavor to describe the Mono Pass in particular, which may, I think, be regarded as a fair example of the higher alpine passes in general.

The main portion of the Mono Pass is formed

by Bloody Cañon, which begins at the summit of the range, and runs in a general east-northeasterly direction to the edge of the Mono Plain.

The first white men who forced a way through its somber depths were, as we have seen, eager gold-seekers. But the cañon was known and traveled as a pass by the Indians and mountain animals long before its discovery by white men, as is shown by the numerous tributary trails which come into it from every direction. Its name accords well with the character of the "early times" in California, and may perhaps have been suggested by the predominant color of the metamorphic slates in which it is in great part eroded; or more probably by blood-stains made by the unfortunate animals which were compelled to slip and shuffle awkwardly over its rough, cutting rocks. I have never known an animal, either mule or horse, to make its way through the cañon, either in going up or down, without losing more or less blood from wounds on the legs. Occasionally one is killed outright — falling headlong and rolling over precipices like a boulder. But such accidents are rarer than from the terrible appearance of the trail one would be led to expect; the more experienced when driven loose find their way over the dangerous places with a caution and sagacity that is truly wonderful. During the gold excitement it was at times a matter of considerable pecuniary importance to force a way through the cañon with pack-trains early in the spring while it was yet heavily blocked with snow; and then the mules with their loads had sometimes to be let down over the steepest drifts and avalanche beds by means of ropes.

A good bridle-path leads from Yosemite through many a grove and meadow up to the head of the cañon, a distance of about thirty miles. Here the scenery undergoes a sudden and startling condensation. Mountains, red, gray, and black, rise close at hand on the right, whitened around their bases with banks of enduring snow; on the left swells the huge red mass of Mount Gibbs, while in front the eye wanders down the shadowy cañon, and out on the warm plain of Mono, where the lake is seen gleaming like a burnished metallic disk, with clusters of lofty volcanic cones to the south of it.

When at length we enter the mountain gateway, the somber rocks seem aware of our presence, and seem to come thronging closer about us. Happily the ouzel and the old familiar robin are here to sing us welcome, and azure daisies beam with trustfulness and sympathy, enabling us to feel something of Nature's love even here, beneath the gaze of her coldest rocks.

The effect of this expressive outspokenness on the part of the cañon-rocks is greatly enhanced by the quiet aspect of the alpine meadows through which we pass just before entering the narrow gateway. The forests in which they lie, and the mountain-tops rising beyond them, seem quiet and tranquil. We catch their restful spirit, yield to the soothing influences of the sunshine, and saunter dreamily on through flowers and bees, scarce touched by a definite thought; then suddenly we find ourselves in the shadowy cañon, closeted with Nature in one of her wildest strongholds.

After the first bewildering impression begins to wear off, we perceive that it is not altogether terrible; for besides the reassuring birds and flowers we discover a chain of shining lakelets hanging down from the very summit of the pass, and linked together by a silvery stream. The highest are set in bleak, rough bowls, scantily fringed with brown and yellow sedges. Winter storms blow snow through the cañon in blinding drifts, and avalanches shoot from the heights. Then are these sparkling tarns filled and buried, leaving not a hint of their existence. In June and July they begin to blink and thaw out like sleepy eyes, the carices thrust up their short brown spikes, the daisies bloom in turn, and the most profoundly buried of them all is at length warmed and summered as if winter were only a dream.

Red Lake is the lowest of the chain, and also the largest. It seems rather dull and forbidding at first sight, lying motionless in its deep, dark bed. The cañon wall rises sheer from the water's edge on the south, but on the opposite side there is sufficient space and sunshine for a sedgy daisy garden, the center of which is brilliantly lighted with lilies, castilleias, larkspurs, and columbines, sheltered from the wind by leafy willows, and forming a most joyful outburst of plant-life keenly emphasized by the chill baldness of the onlooking cliffs.

After indulging here in a dozing, shimmering lake-rest, the happy stream sets forth again, warbling and trilling like an ouzel, ever delightfully confiding, no matter how dark the way; leaping, gliding, hither, thither, clear or foaming: manifesting the beauty of its wildness in every sound and gesture.

One of its most beautiful developments is the Diamond Cascade, situated a short distance below Red Lake. Here the tense, crystalline water is first dashed into coarse, granular spray mixed with dusty foam, and then divided into a diamond pattern by following the diagonal cleavage-joints that intersect the face of the precipice over which it pours. Viewed in front, it resembles a strip of embroidery of definite pattern, varying through the seasons with the temperature and the volume of water. Scarce a flower may be seen along its snowy border. A few bent pines look on from a distance, and small fringes of cassiope and rock-ferns are growing in fissures near the head, but these are so lowly and undemonstrative that only the attentive observer will be likely to notice them.

On the north wall of the cañon, a little below the Diamond Cascade, a glittering side stream makes its appearance, seeming to leap directly out of the sky. It first resembles a crinkled ribbon of silver hanging loosely down the wall, but grows wider as it descends, and dashes the dull rock with foam. A long rough talus curves up against this part of the cliff, overgrown with snow-pressed willows, in which the fall disappears with many an eager surge and swirl and plashing leap, finally beating its way down to its confluence with the main cañon stream.

RANCHERIA FALLS, HETCH-HETCHY VALLEY.

Below this point the climate is no longer arctic. Butterflies become larger and more abundant, grasses with imposing spread of panicle wave above your shoulders, and the summery drone of the bumblebee thickens the air. The Dwarf Pine, the tree-mountaineer that climbs highest and braves the coldest blasts, is found scattered in stormbeaten clumps from the summit of the pass about half-way down the cañon. Here it is succeeded by the hardy Two-leaved Pine, which is speedily joined by the taller Yellow and Mountain Pines. These, with the burly juniper, and shimmering aspen, rapidly grow larger as the sunshine becomes richer, forming groves that block the view; or they stand more apart here and there in picturesque groups, that make beautiful and obvious harmony with the rocks and with one another. Blooming underbrush becomes abundant,— azalea, spiræa, and the brier-rose weaving fringes for the streams, and shaggy rugs to relieve the stern, unflinching rock-bosses.

Through this delightful wilderness, Cañon Creek roves without any constraining channel, throbbing and wavering; now in sunshine, now in thoughtful shade; falling, swirling, flashing from side to side in weariless exuberance of energy. A glorious milky way of cascades is thus developed, of which Bower Cascade, though one of the smallest, is perhaps the most beautiful of them all. It is situated in the lower region of the pass, just where the sunshine begins to mellow between the cold and warm climates. Here the glad creek, grown strong with tribute gathered from many a snowy fountain on the heights, sings richer strains, and becomes more human and lovable at every step. Now you may by its side find the rose and homely yarrow, and small meadows full of bees and clover. At the head of a low-browed rock, luxuriant dogwood bushes and willows arch over from bank to bank, embowering the stream with their leafy branches; and drooping plumes, kept in motion by the current, fringe the brow of the cascade in front. From this leafy covert the stream leaps out into the light in a fluted curve thick sown with sparkling crystals, and falls into a pool filled with brown boulders, out of which it creeps gray with foam-bells and disappears in a tangle of verdure like that from which it came.

Hence, to the foot of the cañon, the metamorphic slates give place to granite, whose nobler sculpture calls forth expressions of corresponding beauty from the stream in passing over it,— bright trills of rapids, booming notes of falls, solemn hushes of smooth-gliding sheets, all chanting and blending in glorious harmony. When, at length, its impetuous alpine life is done, it slips through a meadow with scarce an audible whisper, and falls asleep in Moraine Lake.

This water-bed is one of the finest I ever saw. Evergreens wave soothingly about it, and the breath of flowers floats over it like incense. Here our blessed stream rests from its rocky wanderings, all its mountaineering done,— no more foaming rock-leaping, no more wild, exulting song. It falls into a smooth, glassy sleep, stirred only by the night-wind, which, coming down the cañon, makes it croon and mutter in ripples along its broidered shores.

Leaving the lake, it glides quietly through the rushes, destined never more to touch the living rock. Henceforth its path lies through ancient moraines and reaches of ashy sage-plain, which nowhere afford rocks suitable for the development of cascades or sheer falls. Yet this beauty of maturity, though less striking, is of a still higher order, enticing us lovingly on through gentian meadows and groves of rustling aspen to Lake Mono, where, spirit-like, our happy stream vanishes in vapor, and floats free again in the sky.

Bloody Cañon, like every other in the Sierra, was recently occupied by a glacier, which derived its fountain snows from the adjacent summits, and descended into Mono Lake, at a time when its waters stood at a much higher level than now. The principal characters in which the history of the ancient glaciers is preserved are displayed here in marvelous freshness and simplicity, furnishing the student with extraordinary advantages for the acquisition of knowledge of this sort. The most striking passages are polished and striated surfaces, which in many places reflect the rays of the sun like smooth water. The dam of Red Lake is an elegantly modeled rib of metamorphic slate, brought into relief because of its superior strength, and because of the greater intensity of the glacial erosion of the rock immediately above it, caused by a steeply inclined tributary glacier, which entered the main trunk with a heavy down-thrust at the head of the lake.

Moraine Lake furnishes an equally interesting example of a basin formed wholly, or in part, by a terminal moraine dam curved across the path of a stream between two lateral moraines.

At Moraine Lake the cañon proper terminates, although apparently continued by the two lateral moraines of the vanished glacier. These moraines are about 300 feet high, and extend unbrokenly from the sides of the cañon into the plain, a distance of about five miles, curving and tapering in beautiful lines. Their sunward sides are gardens, their shady sides are groves; the former devoted chiefly to eriogonæ, compositæ, and graminæ; a square rod containing five or six profusely flowered eriogonums of several species, about the same number of bahia and linosyris, and a few grass tufts; each species being planted trimly apart, with bare gravel between, as if cultivated artificially.

My first visit to Bloody Cañon was made in the summer of 1869, under circumstances well calculated to heighten the impressions that are the peculiar offspring of mountains. I came from the blooming tangles of Florida, and waded out into the plant-gold of the great valley of California, when its flora was as yet untrodden. Never before had I beheld congregations of social flowers half so extensive or half so glorious. Golden compositæ covered all the ground from the Coast Range to the Sierra like a stratum of curdled sunshine, in which I reveled for weeks, watching the rising and setting of their innumerable suns; then I gave myself up to be borne forward on the crest of the summer wave that sweeps annually up the Sierra and spends itself on the snowy summits.

At the Big Tuolumne Meadows I remained more
than a month, sketching, botanizing, and climbing
among the surrounding mountains.   The mountaineer with whom I then happened to be camping
was one of those remarkable men one so frequently
meets in California, the hard angles and bosses of
whose characters have been brought into relief by
the grinding excitements of the gold period, until
they resemble glacial landscapes.   But at this late
day, my friend's activities had subsided, and his
craving for rest caused him to become a gentle
shepherd and literally to lie down with the lamb.
Recognizing the unsatisfiable longings of my
Scotch Highland instincts, he threw out some hints
concerning Bloody Cañon, and advised me to explore it.   "I have never seen it myself," he said,
"for I never was so unfortunate as to pass that
way.   But I have heard many a strange story about
it, and I warrant you will at least find it wild
enough."
Then of course I made haste to see it.   Early
next morning I made up a bundle of bread, tied my
note-book to my belt, and strode away in the bracing air, full of eager, indefinite hope.   The plushy
lawns that lay in my path served to soothe my morning haste.   The sod in many places was starred with
daisies and blue gentians, over which I lingered.
I traced the paths of the ancient glaciers over many
a shining pavement, and marked the gaps in the
upper forests that told the power of the winter avalanches.   Climbing higher, I saw for the first time
the gradual dwarfing of the pines in compliance
with climate, and on the summit discovered creeping mats of the arctic willow overgrown with silky

catkins, and patches of the dwarf vaccinium with its round flowers sprinkled in the grass like purple hail; while in every direction the landscape stretched sublimely away in fresh wildness—a manuscript written by the hand of Nature alone.

At length, as I entered the pass, the huge rocks began to close around in all their wild, mysterious impressiveness, when suddenly, as I was gazing eagerly about me, a drove of gray hairy beings came in sight, lumbering toward me with a kind of boneless, wallowing motion like bears.

I never turn back, though often so inclined, and in this particular instance, amid such surroundings, everything seemed singularly unfavorable for the calm acceptance of so grim a company. Suppressing my fears, I soon discovered that although as hairy as bears and as crooked as summit pines, the strange creatures were sufficiently erect to belong to our own species. They proved to be nothing more formidable than Mono Indians dressed in the skins of sage-rabbits. Both the men and the women begged persistently for whisky and tobacco, and seemed so accustomed to denials that I found it impossible to convince them that I had none to give. Excepting the names of these two products of civilization, they seemed to understand not a word of English; but I afterward learned that they were on their way to Yosemite Valley to feast awhile on trout and procure a load of acorns to carry back through the pass to their huts on the shore of Mono Lake.

Occasionally a good countenance may be seen among the Mono Indians, but these, the first specimens I had seen, were mostly ugly, and some of them altogether hideous. The dirt on their faces was fairly stratified, and seemed so ancient and so undisturbed it might almost possess a geological significance. The older faces were, moreover, strangely blurred and divided into sections by furrows that looked like the cleavage-joints of rocks, suggesting exposure on the mountains in a castaway condition for ages. Somehow they seemed to have no right place in the landscape, and I was glad to see them fading out of sight down the pass.

Then came evening, and the somber cliffs were inspired with the ineffable beauty of the alpenglow. A solemn calm fell upon everything. All the lower portion of the cañon was in gloaming shadow, and I crept into a hollow near one of the upper lakelets to smooth the ground in a sheltered nook for a bed. When the short twilight faded, I kindled a sunny fire, made a cup of tea, and lay down to rest and look at the stars. Soon the night-wind began to flow and pour in torrents among the jagged peaks, mingling strange tones with those of the waterfalls sounding far below; and as I drifted toward sleep I began to experience an uncomfortable feeling of nearness to the furred Monos. Then the full moon looked down over the edge of the cañon wall, her countenance seemingly filled with intense concern, and apparently so near as to produce a startling effect as if she had entered my bedroom, forgetting all the world, to gaze on me alone.

The night was full of strange sounds, and I gladly welcomed the morning. Breakfast was soon done, and I set forth in the exhilarating freshness of the new day, rejoicing in the abundance of pure wildness so close about me.   The stupendous rocks, hacked and scarred with centuries of storms, stood sharply out in the thin early light, while down in the bottom of the cañon grooved and polished bosses heaved and glistened like swelling sea-waves, telling a grand old story of the ancient glacier that poured its crushing floods above them.

Here for the first time I met the arctic daisies in all their perfection of purity and spirituality,— gentle mountaineers face to face with the stormy sky, kept safe and warm by a thousand miracles.   I leaped lightly from rock to rock, glorying in the eternal freshness and sufficiency of Nature, and in the ineffable tenderness with which she nurtures her mountain darlings in the very fountains of storms.   Fresh beauty appeared at every step, delicate rock-ferns, and groups of the fairest flowers. Now another lake came to view, now a waterfall. Never fell light in brighter spangles, never fell water in whiter foam.   I seemed to float through the cañon enchanted, feeling nothing of its roughness, and was out in the Mono levels before I was aware.

Looking back from the shore of Moraine Lake, my morning ramble seemed all a dream.   There curved Bloody Cañon, a mere glacial furrow 2000 feet deep, with smooth rocks projecting from the sides and braided together in the middle, like bulging, swelling muscles.   Here the lilies were higher than my head, and the sunshine was warm enough for palms.   Yet the snow around the arctic willows was plainly visible only four miles away, and between were narrow specimen zones of all the principal climates of the globe.

On the bank of a small brook that comes gurgling down the side of the left lateral moraine, I found a camp-fire still burning, which no doubt belonged to the gray Indians I had met on the summit, and I listened instinctively and moved cautiously forward, half expecting to see some of their grim faces peering out of the bushes.

Passing on toward the open plain, I noticed three well-defined terminal moraines curved gracefully across the cañon stream, and joined by long splices to the two noble laterals. These mark the halting-places of the vanished glacier when it was retreating into its summit shadows on the breaking-up of the glacial winter.

Five miles below the foot of Moraine Lake, just where the lateral moraines lose themselves in the plain, there was a field of wild rye, growing in magnificent waving bunches six to eight feet high, bearing heads from six to twelve inches long. Rubbing out some of the grains, I found them about five eighths of an inch long, dark-colored, and sweet. Indian women were gathering it in baskets, bending down large handfuls, beating it out, and fanning it in the wind. They were quite picturesque, coming through the rye, as one caught glimpses of them here and there, in winding lanes and openings, with splendid tufts arching above their heads, while their incessant chat and laughter showed their heedless joy.

Like the rye-field, I found the so-called desert of Mono blooming in a high state of natural cultivation with the wild rose, cherry, aster, and the delicate abronia; also innumerable gilias, phloxes, poppies, and bush-compositæ. I observed their gestures and the various expressions of their corollas, inquiring how they could be so fresh and beautiful out in this volcanic desert. They told as happy a life as any plant-company I ever met, and seemed to enjoy even the hot sand and the wind.

But the vegetation of the pass has been in great part destroyed, and the same may be said of all the more accessible passes throughout the range. Immense numbers of starving sheep and cattle have been driven through them into Nevada, trampling the wild gardens and meadows almost out of existence. The lofty walls are untouched by any foot, and the falls sing on unchanged; but the sight of crushed flowers and stripped, bitten bushes goes far toward destroying the charm of wildness.

The cañon should be seen in winter. A good, strong traveler, who knows the way and the weather, might easily make a safe excursion through it from Yosemite Valley on snow-shoes during some tranquil time, when the storms are hushed. The lakes and falls would be buried then; but so, also, would be the traces of destructive feet, while the views of the mountains in their winter garb, and the ride at lightning speed down the pass between the snowy walls, would be truly glorious.

VIEW OF THE MONO PLAIN FROM THE FOOT OF BLOODY CAÑON.

7

CHAPTER VI

THE GLACIER LAKES

AMONG the many unlooked-for treasures that are bound up and hidden away in the depths of Sierra solitudes, none more surely charm and surprise all kinds of travelers than the glacier lakes. The forests and the glaciers and the snowy fountains of the streams advertise their wealth in a more or less telling manner even in the distance, but nothing is seen of the lakes until we have climbed above them. All the upper branches of the rivers are fairly laden with lakes, like orchard trees with fruit. They lie embosomed in the deep woods, down in the grovy bottoms of cañons, high on bald table-lands, and around the feet of the icy peaks, mirroring back their wild beauty over and over again. Some conception of their lavish abundance may be made from the fact that, from one standpoint on the summit of Red Mountain, a day's journey to the east of Yosemite Valley, no fewer than forty-two are displayed within a radius of ten miles. The whole number in the Sierra can hardly be less than fifteen hundred, not counting the smaller pools and tarns, which are innumerable. Perhaps two thirds or more lie on the western flank of the range, and all are restricted to the alpine and subalpine regions. At the close of the last glacial period, the middle and foot-hill regions also abounded in lakes, all of which have long since vanished as completely as the magnificent ancient glaciers that brought them into existence.

Though the eastern flank of the range is excessively steep, we find lakes pretty regularly distributed throughout even the most precipitous portions. They are mostly found in the upper branches of the cañons, and in the glacial amphitheaters around the peaks.

Occasionally long, narrow specimens occur upon the steep sides of dividing ridges, their basins swung lengthwise like hammocks, and very rarely one is found lying so exactly on the summit of the range at the head of some pass that its waters are discharged down both flanks when the snow is melting fast. But, however situated, they soon cease to form surprises to the studious mountaineer; for, like all the love-work of Nature, they are harmoniously related to one another, and to all the other features of the mountains. It is easy, therefore, to find the bright lake-eyes in the roughest and most ungovernable-looking topography of any landscape countenance. Even in the lower regions, where they have been closed for many a century, their rocky orbits are still discernible, filled in with the detritus of flood and avalanche. A beautiful system of grouping in correspondence with the glacial fountains is soon perceived; also their extension in the direction of the trends of the ancient glaciers; and in general their dependence as to form, size, and position upon the character of the rocks in which their basins have been eroded, and the quantity and direction of application of the glacial force expended upon each basin.

In the upper cañons we usually find them in pretty regular succession, strung together like beads on the bright ribbons of their feeding-streams, which pour, white and gray with foam and spray, from one to the other, their perfect mirror stillness making impressive contrasts with the grand blare and glare of the connecting cataracts. In Lake Hollow, on the north side of the Hoffman spur, immediately above the great Tuolumne cañon, there are ten lovely lakelets lying near together in one general hollow, like eggs in a nest. Seen from above, in a general view, feathered with Hemlock Spruce, and fringed with sedge, they seem to me the most singularly beautiful and interestingly located lake-cluster I have ever yet discovered.

Lake Tahoe, 22 miles long by about 10 wide, and from 500 to over 1600 feet in depth, is the largest of all the Sierra lakes. It lies just beyond the northern limit of the higher portion of the range between the main axis and a spur that puts out on the east side from near the head of the Carson River. Its forested shores go curving in and out around many an emerald bay and pine-crowned promontory, and its waters are everywhere as keenly pure as any to be found among the highest mountains.

Donner Lake, rendered memorable by the terrible fate of the Donner party, is about three miles long, and lies about ten miles to the north of Tahoe, at the head of one of the tributaries of the Truckee. A few miles farther north lies Lake Independence, about the same size as Donner. But far the greater number of the lakes lie much higher and are quite small, few of them exceeding a mile in length, most of them less than half a mile.

Along the lower edge of the lake-belt, the smallest have disappeared by the filling-in of their basins, leaving only those of considerable size. But all along the upper freshly glaciated margin of the lake-bearing zone, every hollow, however small, lying within reach of any portion of the close network of streams, contains a bright, brimming pool; so that the landscape viewed from the mountain-tops seems to be sown broadcast with them. Many of the larger lakes are encircled with smaller ones like central gems girdled with sparkling brilliants. In general, however, there is no marked dividing line as to size. In order, therefore, to prevent confusion, I would state here that in giving numbers, I include none less than 500 yards in circumference.

In the basin of the Merced River, I counted 131, of which 111 are upon the tributaries that fall so grandly into Yosemite Valley. Pohono Creek, which forms the fall of that name, takes its rise in a beautiful lake, lying beneath the shadow of a lofty granite spur that puts out from Buena Vista peak. This is now the only lake left in the whole Pohono Basin. The Illilouette has sixteen, the Nevada no fewer than sixty-seven, the Tenaya eight, Hoffmann Creek five, and Yosemite Creek fourteen. There are but two other lake-bearing affluents of the Merced, viz., the South Fork with fifteen, and Cascade Creek with five, both of which unite with the main trunk below Yosemite.

The Merced River, as a whole, is remarkably like an elm-tree, and it requires but little effort on the part of the imagination to picture it standing upright, with all its lakes hanging upon its spreading branches, the topmost eighty miles in height. Now add all the other lake-bearing rivers of the Sierra, each in its place, and you will have a truly glorious spectacle,— an avenue the length and width of the

LAKE TENAYA, ONE OF THE YOSEMITE FOUNTAINS.

range; the long, slender, gray shafts of the main trunks, the milky way of arching branches, and the silvery lakes, all clearly defined and shining on the sky. How excitedly such an addition to the scenery would be gazed at! Yet these lakeful rivers are still more excitingly beautiful and impressive in their natural positions to those who have the eyes to see them as they lie imbedded in their meadows and forests and glacier-sculptured rocks.

When a mountain lake is born,—when, like a young eye, it first opens to the light,—it is an irregular, expressionless crescent, inclosed in banks of rock and ice,—bare, glaciated rock on the lower side, the rugged snout of a glacier on the upper. In this condition it remains for many a year, until at length, toward the end of some auspicious cluster of seasons, the glacier recedes beyond the upper margin of the basin, leaving it open from shore to shore for the first time, thousands of years after its conception beneath the glacier that excavated its basin. The landscape, cold and bare, is reflected in its pure depths; the winds ruffle its glassy surface, and the sun fills it with throbbing spangles, while its waves begin to lap and murmur around its leafless shores,— sun-spangles during the day and reflected stars at night its only flowers, the winds and the snow its only visitors. Meanwhile, the glacier continues to recede, and numerous rills, still younger than the lake itself, bring down glacier-mud, sand-grains, and pebbles, giving rise to margin-rings and plats of soil. To these fresh soil-beds come many a waiting plant. First, a hardy carex with arching leaves and a spike of brown flowers; then, as the seasons grow warmer, and the soil-beds deeper and wider, other sedges take their appointed places, and these are joined by blue gentians, daisies, dodecatheons, violets, honeyworts, and many a lowly moss. Shrubs also hasten in time to the new gardens,—kalmia with its glossy leaves and purple flowers, the arctic willow, making soft woven carpets, together with the heathy bryanthus and cassiope, the fairest and dearest of them all. Insects now enrich the air, frogs pipe cheerily in the shallows, soon followed by the ouzel, which is the first bird to visit a glacier lake, as the sedge is the first of plants.

So the young lake grows in beauty, becoming more and more humanly lovable from century to century. Groves of aspen spring up, and hardy pines, and the Hemlock Spruce, until it is richly overshadowed and embowered. But while its shores are being enriched, the soil-beds creep out with incessant growth, contracting its area, while the lighter mud-particles deposited on the bottom cause it to grow constantly shallower, until at length the last remnant of the lake vanishes,—closed forever in ripe and natural old age. And now its feeding-stream goes winding on without halting through the new gardens and groves that have taken its place.

The length of the life of any lake depends ordinarily upon the capacity of its basin, as compared with the carrying power of the streams that flow into it, the character of the rocks over which these streams flow, and the relative position of the lake toward other lakes. In a series whose basins lie in the same cañon, and are fed by one and the same main stream, the uppermost will, of course, vanish first unless some other lake-filling agent comes in to modify the result; because at first it receives nearly all of the sediments that the stream brings down, only the finest of the mud-particles being carried through the highest of the series to the next below. Then the next higher, and the next would be successively filled, and the lowest would be the last to vanish. But this simplicity as to duration is broken in upon in various ways, chiefly through the action of side-streams that enter the lower lakes direct. For, notwithstanding many of these side tributaries are quite short, and, during late summer, feeble, they all become powerful torrents in springtime when the snow is melting, and carry not only sand and pine-needles, but large trunks and boulders tons in weight, sweeping them down their steeply inclined channels and into the lake basins with astounding energy. Many of these side affluents also have the advantage of access to the main lateral moraines of the vanished glacier that occupied the cañon, and upon these they draw for lake-filling material, while the main trunk stream flows mostly over clean glacier pavements, where but little moraine matter is ever left for them to carry. Thus a small rapid stream with abundance of loose transportable material within its reach may fill up an extensive basin in a few centuries, while a large perennial trunk stream, flowing over clean, enduring pavements, though ordinarily a hundred times larger, may not fill a smaller basin in thousands of years.

The comparative influence of great and small streams as lake-fillers is strikingly illustrated in Yosemite Valley, through which the Merced flows. The bottom of the valley is now composed of level meadow-lands and dry, sloping soil-beds planted

with oak and pine, but it was once a lake stretching from wall to wall and nearly from one end of
the valley to the other, forming one of the most
beautiful cliff-bound sheets of water that ever
existed in the Sierra. And though never perhaps
seen by human eye, it was but yesterday, geologically speaking, since it disappeared, and the traces
of its existence are still so fresh, it may easily be
restored to the eye of imagination and viewed in
all its grandeur, about as truly and vividly as if
actually before us. Now we find that the detritus
which fills this magnificent basin was not brought
down from the distant mountains by the main
streams that converge here to form the river, however powerful and available for the purpose at first
sight they appear; but almost wholly by the small
local tributaries, such as those of Indian Cañon,
the Sentinel, and the Three Brothers, and by a
few small residual glaciers which lingered in the
shadows of the walls long after the main trunk
glacier had receded beyond the head of the valley.

Had the glaciers that once covered the range
been melted at once, leaving the entire surface
bare from top to bottom simultaneously, then of
course all the lakes would have come into existence
at the same time, and the highest, other circumstances being equal, would, as we have seen, be
the first to vanish. But because they melted gradually from the foot of the range upward, the lower
lakes were the first to see the light and the first
to be obliterated. Therefore, instead of finding the
lakes of the present day at the foot of the range, we
find them at the top. Most of the lower lakes van-

ished thousands of years before those now brightening the alpine landscapes were born. And in
general, owing to the deliberation of the upward

THE DEATH OF A LAKE.

retreat of the glaciers, the lowest of the existing
lakes are also the oldest, a gradual transition being
apparent throughout the entire belt, from the older,
forested, meadow-rimmed and contracted forms all

the way up to those that are new born, lying bare and meadowless among the highest peaks.

A few small lakes unfortunately situated are extinguished suddenly by a single swoop of an avalanche, carrying down immense numbers of trees, together with the soil they were growing upon. Others are obliterated by land-slips, earthquake taluses, etc., but these lake-deaths compared with those resulting from the deliberate and incessant deposition of sediments, may be termed accidental. Their fate is like that of trees struck by lightning.

The lake-line is of course still rising, its present elevation being about 8000 feet above sea-level; somewhat higher than this toward the southern extremity of the range, lower toward the northern, on account of the difference in time of the withdrawal of the glaciers, due to difference in climate. Specimens occur here and there considerably below this limit, in basins specially protected from inwashing detritus, or exceptional in size. These, however, are not sufficiently numerous to make any marked irregularity in the line. The highest I have yet found lies at an elevation of about 12,000 feet, in a glacier womb, at the foot of one of the highest of the summit peaks, a few miles to the north of Mount Ritter. The basins of perhaps twenty-five or thirty are still in process of formation beneath the few lingering glaciers, but by the time they are born, an equal or greater number will probably have died. Since the beginning of the close of the ice-period the whole number in the range has perhaps never been greater than at present.

A rough approximation to the average duration

of these mountain lakes may be made from data already suggested, but I cannot stop here to present the subject in detail. I must also forego, in the mean time, the pleasure of a full discussion of the interesting question of lake-basin formation, for which fine, clear, demonstrative material abounds in these mountains. In addition to what has been already given on the subject, I will only make this one statement. Every lake in the Sierra is a glacier lake. Their basins were not merely remodeled and scoured out by this mighty agent, but in the first place were eroded from the solid.

I must now make haste to give some nearer views of representative specimens lying at different elevations on the main lake-belt, confining myself to descriptions of the features most characteristic of each.

SHADOW LAKE

This is a fine specimen of the oldest and lowest of the existing lakes. It lies about eight miles above Yosemite Valley, on the main branch of the Merced, at an elevation of about 7350 feet above the sea; and is everywhere so securely cliff-bound that without artificial trails only wild animals can get down to its rocky shores from any direction. Its original length was about a mile and a half; now it is only half a mile in length by about a fourth of a mile in width, and over the lowest portion of the basin ninety-eight feet deep. Its crystal waters are clasped around on the north and south by majestic granite walls sculptured in true Yosemitic style into domes,

gables, and battlemented headlands, which on the
south come plunging down sheer into deep water,
from a height of from 1500 to 2000 feet. The South
Lyell glacier eroded this magnificent basin out of
solid porphyritic granite while forcing its way westward from the summit fountains toward Yosemite,
and the exposed rocks around the shores, and the
projecting bosses of the walls, ground and burnished
beneath the vast ice-flood, still glow with silvery
radiance, notwithstanding the innumerable corroding storms that have fallen upon them.   The general conformation of the basin, as well as the moraines laid along the top of the walls, and the
grooves and scratches on the bottom and sides,
indicate in the most unmistakable manner the direction pursued by this mighty ice-river, its great
depth, and the tremendous energy it exerted in
thrusting itself into and out of the basin; bearing
down with superior pressure upon this portion of
its channel, because of the greater declivity, consequently eroding it deeper than the other portions
about it, and producing the lake-bowl as the necessary result.

With these magnificent ice-characters so vividly
before us it is not easy to realize that the old glacier
that made them vanished tens of centuries ago; for,
excepting the vegetation that has sprung up, and
the changes effected by an earthquake that hurled
rock-avalanches from the weaker headlands, the
basin as a whole presents the same appearance
that it did when first brought to light. The lake
itself, however, has undergone marked changes;
one sees at a glance that it is growing old. More

SHADOW LAKE (MERCED LAKE), YOSEMITE NATIONAL PARK.

than two thirds of its original area is now dry land,
covered with meadow-grasses and groves of pine
and fir, and the level bed of alluvium stretching
across from wall to wall at the head is evidently
growing out all along its lakeward margin, and will
at length close the lake forever.

Every lover of fine wildness would delight to
saunter on a summer day through the flowery
groves now occupying the filled-up portion of the
basin. The curving shore is clearly traced by a
ribbon of white sand upon which the ripples play;
then comes a belt of broad-leafed sedges, interrupted here and there by impenetrable tangles of
willows; beyond this there are groves of trembling
aspen; then a dark, shadowy belt of Two-leaved
Pine, with here and there a round carex meadow
ensconced nest-like in its midst; and lastly, a narrow outer margin of majestic Silver Fir 200 feet
high. The ground beneath the trees is covered
with a luxuriant crop of grasses, chiefly triticum,
bromus, and calamagrostis, with purple spikes and
panicles arching to one's shoulders; while the open
meadow patches glow throughout the summer with
showy flowers,—heleniums, goldenrods, erigerons,
lupines, castilleias, and lilies, and form favorite hid-
ing- and feeding-grounds for bears and deer.

The rugged south wall is feathered darkly along
the top with an imposing array of spirey Silver Firs,
while the rifted precipices all the way down to the
water's edge are adorned with picturesque old junipers, their cinnamon-colored bark showing finely
upon the neutral gray of the granite. These, with
a few venturesome Dwarf Pines and Spruces, lean

out over fissured ribs and tablets, or stand erect back in shadowy niches, in an indescribably wild and fearless manner. Moreover, the white-flowered Douglas spiræa and dwarf evergreen oak form graceful fringes along the narrower seams, wherever the slightest hold can be effected. Rock-ferns, too, are here, such as allosorus, pellæa, and cheilanthes, making handsome rosettes on the drier fissures; and the delicate maidenhair, cistoperis, and woodsia hide back in mossy grottoes, moistened by some trickling rill; and then the orange wall-flower holds up its showy panicles here and there in the sunshine, and bahia makes bosses of gold. But, notwithstanding all this plant beauty, the general impression in looking across the lake is of stern, unflinching rockiness; the ferns and flowers are scarcely seen, and not one fiftieth of the whole surface is screened with plant life.

The sunnier north wall is more varied in sculpture, but the general tone is the same. A few headlands, flat-topped and soil-covered, support clumps of cedar and pine; and up-curving tangles of chinquapin and live-oak, growing on rough earthquake taluses, girdle their bases. Small streams come cascading down between them, their foaming margins brightened with gay primulas, gilias, and mimuluses. And close along the shore on this side there is a strip of rocky meadow enameled with buttercups, daisies, and white violets, and the purple-topped grasses out on its beveled border dip their leaves into the water.

The lower edge of the basin is a dam-like swell of solid granite, heavily abraded by the old glacier, but scarce at all cut into as yet by the outflowing stream, though it has flowed on unceasingly since the lake came into existence.

As soon as the stream is fairly over the lake-lip it breaks into cascades, never for a moment halting, and scarce abating one jot of its glad energy, until it reaches the next filled-up basin, a mile below. Then swirling and curving drowsily through meadow and grove, it breaks forth anew into gray rapids and falls, leaping and gliding in glorious exuberance of wild bound and dance down into another and yet another filled-up lake basin. Then, after a long rest in the levels of Little Yosemite, it makes its grandest display in the famous Nevada Fall. Out of the clouds of spray at the foot of the fall the battered, roaring river gropes its way, makes another mile of cascades and rapids, rests a moment in Emerald Pool, then plunges over the grand cliff of the Vernal Fall, and goes thundering and chafing down a boulder-choked gorge of tremendous depth and wildness into the tranquil reaches of the old Yosemite lake basin.

The color-beauty about Shadow Lake during the Indian summer is much richer than one could hope to find in so young and so glacial a wilderness. Almost every leaf is tinted then, and the golden-rods are in bloom; but most of the color is given by the ripe grasses, willows, and aspens. At the foot of the lake you stand in a trembling aspen grove, every leaf painted like a butterfly, and away to right and left round the shores sweeps a curving ribbon of meadow, red and brown dotted with pale yellow, shading off here and there into hazy

8

purple. The walls, too, are dashed with bits of bright color that gleam out on the neutral granite gray. But neither the walls, nor the margin meadow, nor yet the gay, fluttering grove in which you stand, nor the lake itself, flashing with spangles, can long hold your attention; for at the head of the lake there is a gorgeous mass of orange-yellow, belonging to the main aspen belt of the basin, which seems the very fountain whence all the color below it had flowed, and here your eye is filled and fixed. This glorious mass is about thirty feet high, and extends across the basin nearly from wall to wall. Rich bosses of willow flame in front of it, and from the base of these the brown meadow comes forward to the water's edge, the whole being relieved against the unyielding green of the coniferæ, while thick sun-gold is poured over all.

During these blessed color-days no cloud darkens the sky, the winds are gentle, and the landscape rests, hushed everywhere, and indescribably impressive. A few ducks are usually seen sailing on the lake, apparently more for pleasure than anything else, and the ouzels at the head of the rapids sing always; while robins, grosbeaks, and the Douglas squirrels are busy in the groves, making delightful company, and intensifying the feeling of grateful sequestration without ruffling the deep, hushed calm and peace.

This autumnal mellowness usually lasts until the end of November. Then come days of quite another kind. The winter clouds grow, and bloom, and shed their starry crystals on every leaf and rock, and all the colors vanish like a sunset. The deer gather

VERNAL FALL, YOSEMITE VALLEY.

and hasten down their well-known trails, fearful of being snow-bound. Storm succeeds storm, heaping snow on the cliffs and meadows, and bending the slender pines to the ground in wide arches, one over the other, clustering and interlacing like lodged wheat. Avalanches rush and boom from the shelving heights, piling immense heaps upon the frozen lake, and all the summer glory is buried and lost. Yet in the midst of this hearty winter the sun shines warm at times, calling the Douglas squirrel to frisk in the snowy pines and seek out his hidden stores; and the weather is never so severe as to drive away the grouse and little nut-hatches and chickadees.

Toward May, the lake begins to open. The hot sun sends down innumerable streams over the cliffs, streaking them round and round with foam. The snow slowly vanishes, and the meadows show tintings of green. Then spring comes on apace; flowers and flies enrich the air and the sod, and the deer come back to the upper groves like birds to an old nest.

I first discovered this charming lake in the autumn of 1872, while on my way to the glaciers at the head of the river. It was rejoicing then in its gayest colors, untrodden, hidden in the glorious wildness like unmined gold. Year after year I walked its shores without discovering any other trace of humanity than the remains of an Indian camp-fire, and the thigh-bones of a deer that had been broken to get at the marrow. It lies out of the regular ways of Indians, who love to hunt in more accessible fields adjacent to trails. Their knowledge of deer-haunts had probably enticed them here some hunger-time when they wished to make sure of a feast; for hunting in this lake-hollow is like hunting in a fenced park. I had told the beauty of Shadow Lake only to a few friends, fearing it might come to be trampled and "improved" like Yosemite. On my last visit, as I was sauntering along the shore on the strip of sand between the water and sod, reading the tracks of the wild animals that live here, I was startled by a human track, which I at once saw belonged to some shepherd; for each step was turned out 35° or 40° from the general course pursued, and was also run over in an uncertain sprawling fashion at the heel, while a row of round dots on the right indicated the staff that shepherds carry. None but a shepherd could make such a track, and after tracing it a few minutes I began to fear that he might be seeking pasturage; for what else could he be seeking? Returning from the glaciers shortly afterward, my worst fears were realized. A trail had been made down the mountain-side from the north, and all the gardens and meadows were destroyed by a horde of hoofed locusts, as if swept by a fire. The money-changers were in the temple.

ORANGE LAKE

BESIDES these larger cañon lakes, fed by the main cañon streams, there are many smaller ones lying aloft on the top of rock benches, entirely independent of the general drainage channels, and of course drawing their supplies from a very limited area. Notwithstanding they are mostly small and shallow, owing to their immunity from avalanche detritus and the inwashings of powerful streams, they often endure longer than others many times larger but less favorably situated. When very shallow they become dry toward the end of summer; but because their basins are ground out of seamless stone they suffer no loss save from evaporation alone; and the great depth of snow that falls, lasting into June, makes their dry season short in any case.

Orange Lake is a fair illustration of this bench form. It lies in the middle of a beautiful glacial pavement near the lower margin of the lake-line, about a mile and a half to the northwest of Shadow Lake. It is only about 100 yards in circumference. Next the water there is a girdle of carices with wide overarching leaves, then in regular order a shaggy ruff of huckleberry bushes, a zone of willows with here and there a bush of the Mountain Ash, then a zone of aspens with a few pines around the outside. These zones are of course concentric, and together form a wall beyond which the naked ice-burnished granite stretches away in every direction, leaving it conspicuously relieved, like a bunch of palms in a desert.

In autumn, when the colors are ripe, the whole circular grove, at a little distance, looks like a big handful of flowers set in a cup to be kept fresh — a tuft of goldenrods. Its feeding-streams are exceedingly beautiful, notwithstanding their inconstancy and extreme shallowness. They have no channel whatever, and consequently are left free to spread in thin sheets upon the shining granite and wander at will. In many places the current is less than a fourth of an inch deep, and flows with so little friction it is scarcely visible. Sometimes there is not a single foam-bell, or drifting pine-needle, or irregularity of any sort to manifest its motion. Yet when observed narrowly it is seen to form a web of gliding lacework exquisitely woven, giving beautiful reflections from its minute curving ripples and eddies, and differing from the water-laces of large cascades in being everywhere transparent. In spring, when the snow is melting, the lake-bowl is brimming full, and sends forth quite a large stream that slips glassily for 200 yards or so, until it comes to an almost vertical precipice 800 feet high, down which it plunges in a fine cataract; then it gathers its scattered waters and goes smoothly over folds of gently dipping granite to its confluence with the main cañon stream. During the greater portion of the year, however, not a single water sound will you hear either at head or foot of the lake, not even the whispered lappings of ripple-waves along the shore; for the winds are fenced out. But the deep mountain silence is sweetened now and then by birds that stop here to rest and drink on their way across the cañon.

LAKE STARR KING

A BEAUTIFUL variety of the bench-top lakes occurs just where the great lateral moraines of the main glaciers have been shoved forward in outswelling concentric rings by small residual tributary glaciers.

LAKE STARR KING.

Instead of being encompassed by a narrow ring of trees like Orange Lake, these lie embosomed in dense moraine woods, so dense that in seeking them you may pass them by again and again, although you may know nearly where they lie concealed.

Lake Starr King, lying to the north of the cone of that name, above the Little Yosemite Valley, is a fine specimen of this variety. The ouzels pass it by, and so do the ducks; they could hardly get into it if they would, without plumping straight down inside the circling trees.

Yet these isolated gems, lying like fallen fruit detached from the branches, are not altogether without inhabitants and joyous, animating visitors. Of course fishes cannot get into them, and this is generally true of nearly every glacier lake in the range, but they are all well stocked with happy frogs. How did the frogs get into them in the first place? Perhaps their sticky spawn was carried in on the feet of ducks or other birds, else their progenitors must have made some exciting excursions through the woods and up the sides of the cañons. Down in the still, pure depths of these hidden lakelets you may also find the larvæ of innumerable insects and a great variety of beetles, while the air above them is thick with humming wings, through the midst of which fly-catchers are constantly darting. And in autumn, when the huckleberries are ripe, bands of robins and grosbeaks come to feast, forming altogether delightful little byworlds for the naturalist.

Pushing our way upward toward the axis of the range, we find lakes in greater and greater abundance, and more youthful in aspect. At an elevation of about 9000 feet above sea-level they seem to have arrived at middle age,— that is, their basins seem to be about half filled with alluvium. Broad sheets of meadow-land are seen extending into them, imperfect and boggy in many places and more nearly level than those of the older lakes below them, and the vegetation of their shores is of course more alpine. Kalmia, ledum, and cassiope fringe the meadow rocks, while the luxuriant, waving groves, so characteristic of the lower lakes, are represented only by clumps of the Dwarf Pine and Hemlock Spruce. These, however, are oftentimes very picturesquely grouped on rocky headlands around the outer rim of the meadows, or with still more striking effect crown some rocky islet.

Moreover, from causes that we cannot stop here to explain, the cliffs about these middle-aged lakes are seldom of the massive Yosemite type, but are more broken, and less sheer, and they usually stand back, leaving the shores comparatively free; while the few precipitous rocks that do come forward and plunge directly into deep water are seldom more than three or four hundred feet high.

I have never yet met ducks in any of the lakes of this kind, but the ouzel is never wanting where the feeding-streams are perennial. Wild sheep and deer may occasionally be seen on the meadows, and very rarely a bear. One might camp on the rugged shores of these bright fountains for weeks, without meeting any animal larger than the marmots that burrow beneath glacier boulders along the edges of the meadows.

The highest and youngest of all the lakes lie

nestled in glacier wombs. At first sight, they seem
pictures of pure bloodless desolation, miniature
arctic seas, bound in perpetual ice and snow, and
overshadowed by harsh, gloomy, crumbling precipices. Their waters are keen ultramarine blue in
the deepest parts, lively grass-green toward the shore
shallows and around the edges of the small bergs
usually floating about in them. A few hardy
sedges, frost-pinched every night, are occasionally
found making soft sods along the sun-touched portions of their shores, and when their northern banks
slope openly to the south, and are soil-covered, no
matter how coarsely, they are sure to be brightened
with flowers. One lake in particular now comes to
mind which illustrates the floweriness of the sun-
touched banks of these icy gems. Close up under
the shadow of the Sierra Matterhorn, on the
eastern slope of the range, lies one of the iciest of
these glacier lakes at an elevation of about 12,000
feet. A short, ragged-edged glacier crawls into it
from the south, and on the opposite side it is em-
banked and dammed by a series of concentric terminal moraines, made by the glacier when it entirely filled the basin. Half a mile below lies a
second lake, at a height of 11,500 feet, about as cold
and as pure as a snow-crystal. The waters of the
first come gurgling down into it over and through
the moraine dam, while a second stream pours into
it direct from a glacier that lies to the southeast.
Sheer precipices of crystalline snow rise out of deep
water on the south, keeping perpetual winter on that
side, but there is a fine summery spot on the other,
notwithstanding the lake is only about 300 yards

wide. Here, on August 25, 1873, I found a charming company of flowers, not pinched, crouching dwarfs, scarce able to look up, but warm and juicy, standing erect in rich cheery color and bloom. On a narrow strip of shingle, close to the water's edge, there were a few tufts of carex gone to seed; and a little way back up the rocky bank at the foot of a crumbling wall so inclined as to absorb and radiate as well as reflect a considerable quantity of sun-heat, was the garden, containing a thrifty thicket of Cowania covered with large yellow flowers; several bushes of the alpine ribes with berries nearly ripe and wildly acid; a few handsome grasses belonging to two distinct species, and one goldenrod; a few hairy lupines and radiant spragueas, whose blue and rose-colored flowers were set off to fine advantage amid green carices; and along a narrow seam in the very warmest angle of the wall a perfectly gorgeous fringe of *Epilobium obcordatum* with flowers an inch wide, crowded together in lavish profusion, and colored as royal a purple as ever was worn by any high-bred plant of the tropics; and best of all, and greatest of all, a noble thistle in full bloom, standing erect, head and shoulders above his companions, and thrusting out his lances in sturdy vigor as if growing on a Scottish brae. All this brave warm bloom among the raw stones, right in the face of the onlooking glaciers.

As far as I have been able to find out, these upper lakes are snow-buried in winter to a depth of about thirty-five or forty feet, and those most exposed to avalanches, to a depth of even a hundred feet or more. These last are, of course, nearly lost to the landscape. Some remain buried for years, when the snowfall is exceptionally great, and many open only on one side late in the season. The snow of the closed side is composed of coarse granules compacted and frozen into a firm, faintly stratified mass, like the *névé* of a glacier. The lapping waves of the open portion gradually undermine and cause it to break off in large masses like icebergs, which gives rise to a precipitous front like the discharging wall of a glacier entering the sea. The play of the lights among the crystal angles of these snow-cliffs, the pearly white of the outswelling bosses, the bergs drifting in front, aglow in the sun and edged with green water, and the deep blue disk of the lake itself extending to your feet,—this forms a picture that enriches all your afterlife, and is never forgotten. But however perfect the season and the day, the cold incompleteness of these young lakes is always keenly felt. We approach them with a kind of mean caution, and steal unconfidingly around their crystal shores, dashed and ill at ease, as if expecting to hear some forbidding voice. But the love-songs of the ouzels and the love-looks of the daisies gradually reassure us, and manifest the warm fountain humanity that pervades the coldest and most solitary of them all.

CHAPTER VII

THE GLACIER MEADOWS

AFTER the lakes on the High Sierra come the glacier meadows. They are smooth, level, silky lawns, lying embedded in the upper forests, on the floors of the valleys, and along the broad backs of the main dividing ridges, at a height of about 8000 to 9500 feet above the sea.

They are nearly as level as the lakes whose places they have taken, and present a dry, even surface free from rock-heaps, mossy bogginess, and the frowsy roughness of rank, coarse-leaved, weedy, and shrubby vegetation. The sod is close and fine, and so complete that you cannot see the ground; and at the same time so brightly enameled with flowers and butterflies that it may well be called a garden-meadow, or meadow-garden; for the plushy sod is in many places so crowded with gentians, daisies, ivesias, and various species of orthocarpus that the grass is scarcely noticeable, while in others the flowers are only pricked in here and there singly, or in small ornamental rosettes.

The most influential of the grasses composing the sod is a delicate calamagrostis with fine filiform leaves, and loose, airy panicles that seem to float above the flowery lawn like a purple mist. But, write as I may, I cannot give anything like an adequate idea of the exquisite beauty of these mountain carpets as they lie smoothly outspread in the savage wilderness. What words are fine enough to picture them? to what shall we liken them? The flowery levels of the prairies of the old West, the luxuriant savannahs of the South, and the finest of cultivated meadows are coarse in comparison. One may at first sight compare them with the carefully tended lawns of pleasure-grounds; for they are as free from weeds as they, and as smooth, but here the likeness ends; for these wild lawns, with all their exquisite fineness, have no trace of that painful, licked, snipped, repressed appearance that pleasure-ground lawns are apt to have even when viewed at a distance. And, not to mention the flowers with which they are brightened, their grasses are very much finer both in color and texture, and instead of lying flat and motionless, matted together like a dead green cloth, they respond to the touches of every breeze, rejoicing in pure wildness, blooming and fruiting in the vital light.

Glacier meadows abound throughout all the alpine and subalpine regions of the Sierra in still greater numbers than the lakes. Probably from 2500 to 3000 exist between latitude 36° 30′ and 39°, distributed, of course, like the lakes, in concordance with all the other glacial features of the landscape.

On the head waters of the rivers there are what are called "Big Meadows," usually about from five to ten miles long. These occupy the basins of the ancient ice-seas, where many tributary glaciers came together to form the grand trunks. Most, however, are quite small, averaging perhaps but little more than three fourths of a mile in length.

One of the very finest of the thousands I have enjoyed lies hidden in an extensive forest of the Two-leaved Pine, on the edge of the basin of the ancient Tuolumne Mer de Glace, about eight miles to the west of Mount Dana.

Imagine yourself at the Tuolumne Soda Springs on the bank of the river, a day's journey above Yosemite Valley. You set off northward through a forest that stretches away indefinitely before you, seemingly unbroken by openings of any kind. As soon as you are fairly into the woods, the gray mountain-peaks, with their snowy gorges and hollows, are lost to view. The ground is littered with fallen trunks that lie crossed and recrossed like storm-lodged wheat; and besides this close forest of pines, the rich moraine soil supports a luxuriant growth of ribbon-leaved grasses—bromus, triticum, calamagrostis, agrostis, etc., which rear their handsome spikes and panicles above your waist. Making your way through the fertile wilderness,— finding lively bits of interest now and then in the squirrels and Clark crows, and perchance in a deer or bear,— after the lapse of an hour or two vertical bars of sunshine are seen ahead between the brown shafts of the pines, showing that you are approaching an open space, and then you suddenly emerge from the forest shadows upon a delightful purple lawn lying smooth and free in the light like a lake. This is a glacier meadow. It is about a mile and a half long by a quarter of a mile wide. The trees come pressing forward all around in close serried ranks, planting their feet exactly on its margin, and holding themselves erect, strict and orderly like soldiers on parade; thus bounding the meadow with exquisite precision, yet with free curving lines such as Nature alone can draw. With inexpressible delight you wade out into the grassy sun-lake, feeling yourself contained in one of Nature's most sacred chambers, withdrawn from the sterner influences of the mountains, secure from all intrusion, secure from yourself, free in the universal beauty. And notwithstanding the scene is so impressively spiritual, and you seem dissolved in it, yet everything about you is beating with warm, terrestrial, human love and life delightfully substantial and familiar. The resiny pines are types of health and steadfastness; the robins feeding on the sod belong to the same species you have known since childhood; and surely these daisies, larkspurs, and goldenrods are the very friend-flowers of the old home garden. Bees hum as in a harvest noon, butterflies waver above the flowers, and like them you lave in the vital sunshine, too richly and homogeneously joy-filled to be capable of partial thought. You are all eye, sifted through and through with light and beauty. Sauntering along the brook that meanders silently through the meadow from the east, special flowers call you back to discriminating consciousness. The sod comes curving down to the water's edge, forming bossy outswelling banks, and in some places overlapping countersunk boulders and forming bridges. Here you find mats of the curious dwarf willow scarce an inch high, yet sending up a multitude of gray silky catkins, illumined here and there with the purple cups and bells of bryanthus and vaccinium.

Go where you may, you everywhere find the lawn divinely beautiful, as if Nature had fingered and adjusted every plant this very day. The floating grass panicles are scarcely felt in brushing through their midst, so fine are they, and none of the flowers have tall or rigid stalks. In the brightest places you find three species of gentians with different shades of blue, daisies pure as the sky, silky leaved ivesias with warm yellow flowers, several species of orthocarpus with blunt, bossy spikes, red and purple and yellow; the alpine goldenrod, pentstemon, and clover, fragrant and honeyful, with their colors massed and blended. Parting the grasses and looking more closely you may trace the branching of their shining stems, and note the marvelous beauty of their mist of flowers, the glumes and pales exquisitely penciled, the yellow dangling stamens, and feathery pistils. Beneath the lowest leaves you discover a fairy realm of mosses,—hypnum, dicranum, polytrichum, and many others,—their precious spore-cups poised daintily on polished shafts, curiously hooded, or open, showing the richly ornate peristomas worn like royal crowns. Creeping liverworts are here also in abundance, and several rare species of fungi, exceedingly small, and frail, and delicate, as if made only for beauty. Caterpillars, black beetles, and ants roam the wilds of this lower world, making their way through miniature groves and thickets like bears in a thick wood.

And how rich, too, is the life of the sunny air! Every leaf and flower seems to have its winged

9

representative overhead. Dragon-flies shoot in vigorous zigzags through the dancing swarms, and a
rich profusion of butterflies — the leguminosæ of insects — make a fine addition to the general show.
Many of these last are comparatively small at this
elevation, and as yet almost unknown to science;
but every now and then a familiar vanessa or papilio
comes sailing past. Humming-birds, too, are quite
common here, and the robin is always found along
the margin of the stream, or out in the shallowest
portions of the sod, and sometimes the grouse and
mountain quail, with their broods of precious fluffy
chickens. Swallows skim the grassy lake from end
to end, fly-catchers come and go in fitful flights
from the tops of dead spars, while woodpeckers
swing across from side to side in graceful festoon
curves,— birds, insects, and flowers all in their own
way telling a deep summer joy.

The influences of pure nature seem to be so little
known as yet, that it is generally supposed that
complete pleasure of this kind, permeating one's very
flesh and bones, unfits the student for scientific
pursuits in which cool judgment and observation
are required. But the effect is just the opposite.
Instead of producing a dissipated condition, the
mind is fertilized and stimulated and developed like
sun-fed plants. All that we have seen here enables
us to see with surer vision the fountains among the
summit-peaks to the east whence flowed the glaciers
that ground soil for the surrounding forest; and
down at the foot of the meadow the moraine which
formed the dam which gave rise to the lake that
occupied this basin before the meadow was made;

and around the margin the stones that were shoved back and piled up into a rude wall by the expansion of the lake ice during long bygone winters; and along the sides of the streams the slight hollows of the meadow which mark those portions of the old lake that were the last to vanish.

I would fain ask my readers to linger awhile in this fertile wilderness, to trace its history from its earliest glacial beginnings, and learn what we may of its wild inhabitants and visitors. How happy the birds are all summer and some of them all winter; how the pouched marmots drive tunnels under the snow, and how fine and brave a life the slandered coyote lives here, and the deer and bears! But, knowing well the difference between reading and seeing, I will only ask attention to some brief sketches of its varying aspects as they are presented throughout the more marked seasons of the year.

The summer life we have been depicting lasts with but little abatement until October, when the night frosts begin to sting, bronzing the grasses, and ripening the leaves of the creeping heathworts along the banks of the stream to reddish purple and crimson; while the flowers disappear, all save the goldenrods and a few daisies, that continue to bloom on unscathed until the beginning of snowy winter. In still nights the grass panicles and every leaf and stalk are laden with frost crystals, through which the morning sunbeams sift in ravishing splendor, transforming each to a precious diamond radiating the colors of the rainbow. The brook shallows are plaited across and across with slender lances of ice,

132   THE MOUNTAINS OF CALIFORNIA

but both these and the grass crystals are melted before midday, and, notwithstanding the great elevation of the meadow, the afternoons are still warm enough to revive the chilled butterflies and call them out to enjoy the late-flowering goldenrods. The divine alpenglow flushes the surrounding forest every evening, followed by a crystal night with hosts of lily stars, whose size and brilliancy cannot be conceived by those who have never risen above the lowlands.

Thus come and go the bright sun-days of autumn, not a cloud in the sky, week after week until near December. Then comes a sudden change. Clouds of a peculiar aspect with a slow, crawling gait gather and grow in the azure, throwing out satiny fringes, and becoming gradually darker until every lake-like rift and opening is closed and the whole bent firmament is obscured in equal structureless gloom. Then comes the snow, for the clouds are ripe, the meadows of the sky are in bloom, and shed their radiant blossoms like an orchard in the spring. Lightly, lightly they lodge in the brown grasses and in the tasseled needles of the pines, falling hour after hour, day after day, silently, lovingly,— all the winds hushed,— glancing and circling hither, thither, glinting against one another, rays interlocking in flakes as large as daisies; and then the dry grasses, and the trees, and the stones are all equally abloom again. Thunder-showers occur here during the summer months, and impressive it is to watch the coming of the big transparent drops, each a small world in itself,— one unbroken ocean without islands hurling free through the air like planets through space. But still more impressive to me is the coming of the snow-flowers,— falling stars, winter daisies,— giving bloom to all the ground alike. Raindrops blossom brilliantly in the rainbow, and change to flowers in the sod, but snow comes in full flower direct from the dark, frozen sky.

The later snow-storms are oftentimes accompanied by winds that break up the crystals, when the temperature is low, into single petals and irregular dusty fragments; but there is comparatively little drifting on the meadow, so securely is it embosomed in the woods. From December to May, storm succeeds storm, until the snow is about fifteen or twenty feet deep, but the surface is always as smooth as the breast of a bird.

Hushed now is the life that so late was beating warmly. Most of the birds have gone down below the snow-line, the plants sleep, and all the fly-wings are folded. Yet the sun beams gloriously many a cloudless day in midwinter, casting long lance shadows athwart the dazzling expanse. In June small flecks of the dead, decaying sod begin to appear, gradually widening and uniting with one another, covered with creeping rags of water during the day, and ice by night, looking as hopeless and unvital as crushed rocks just emerging from the darkness of the glacial period. Walk the meadow now! Scarce the memory of a flower will you find. The ground seems twice dead. Nevertheless, the annual resurrection is drawing near. The life-giving sun pours his floods, the last snow-wreath melts, myriads of growing points push eagerly through the steaming mold, the birds come back, new wings fill the air, and fervid summer life comes surging on, seemingly yet more glorious than before.

This is a perfect meadow, and under favorable circumstances exists without manifesting any marked changes for centuries. Nevertheless, soon or late it must inevitably grow old and vanish. During the calm Indian summer, scarce a sand-grain moves around its banks, but in flood-times and storm-times, soil is washed forward upon it and laid in successive sheets around its gently sloping rim, and is gradually extended to the center, making it dryer. Through a considerable period the meadow vegetation is not greatly affected thereby, for it gradually rises with the rising ground, keeping on the surface like water-plants rising on the swell of waves. But at length the elevation of the meadow-land goes on so far as to produce too dry a soil for the specific meadow-plants, when, of course, they have to give up their places to others fitted for the new conditions. The most characteristic of the new-comers at this elevation above the sea are prin-cipally sun-loving gilias, eriogonæ, and compositæ, and finally forest-trees. Henceforward the obscur-ing changes are so manifold that the original lake-meadow can be unveiled and seen only by the geologist.

Generally speaking, glacier lakes vanish more slowly than the meadows that succeed them, be-cause, unless very shallow, a greater quantity of ma-terial is required to fill up their basins and obliterate them than is required to render the surface of the meadow too high and dry for meadow vegetation.

Furthermore, owing to the weathering to which the adjacent rocks are subjected, material of the finer sort, susceptible of transportation by rains and ordinary floods, is more abundant during the meadow period than during the lake period. Yet doubtless many a fine meadow favorably situated exists in almost prime beauty for thousands of years, the process of extinction being exceedingly slow, as we reckon time. This is especially the case with meadows circumstanced like the one we have described — embosomed in deep woods, with the ground rising gently away from it all around, the network of tree-roots in which all the ground is clasped preventing any rapid torrential washing. But, in exceptional cases, beautiful lawns formed with great deliberation are overwhelmed and obliterated at once by the action of land-slips, earthquake avalanches, or extraordinary floods, just as lakes are.

In those glacier meadows that take the places of shallow lakes which have been fed by feeble streams, glacier mud and fine vegetable humus enter largely into the composition of the soil; and on account of the shallowness of this soil, and the seamless, water-tight, undrained condition of the rock-basins, they are usually wet, and therefore occupied by tall grasses and sedges, whose coarse appearance offers a striking contrast to that of the delicate lawn-making kind described above. These shallow-soiled meadows are oftentimes still further roughened and diversified by partially buried moraines and swelling bosses of the bed-rock, which, with the trees and shrubs growing upon them, produce a striking effect

as they stand in relief like islands in the grassy
level, or sweep across in rugged curves from one
forest wall to the other.

Throughout the upper meadow region, wherever
water is sufficiently abundant and low in temperature, in basins secure from flood-washing, handsome
bogs are formed with a deep growth of brown and
yellow sphagnum picturesquely ruffled with patches
of kalmia and ledum which ripen masses of beautiful color in the autumn. Between these cool,
spongy bogs and the dry, flowery meadows there
are many interesting varieties which are graduated
into one another by the varied conditions already
alluded to, forming a series of delightful studies.

HANGING MEADOWS

ANOTHER very well-marked and interesting kind
of meadow, differing greatly both in origin and appearance from the lake-meadows, is found lying
aslant upon moraine-covered hillsides trending in
the direction of greatest declivity, waving up and
down over rock heaps and ledges, like rich green
ribbons brilliantly illumined with tall flowers.
They occur both in the alpine and subalpine regions in considerable numbers, and never fail to
make telling features in the landscape. They are
often a mile or more in length, but never very wide
—usually from thirty to fifty yards. When the
mountain or cañon side on which they lie dips at
the required angle, and other conditions are at the
same time favorable, they extend from above the

timber line to the bottom of a cañon or lake basin, descending in fine, fluent lines like cascades, breaking here and there into a kind of spray on large boulders, or dividing and flowing around on either side of some projecting islet. Sometimes a noisy stream goes brawling down through them, and again, scarcely a drop of water is in sight. They owe their existence, however, to streams, whether visible or invisible, the wildest specimens being found where some perennial fountain, as a glacier or snowbank or moraine spring sends down its waters across a rough sheet of soil in a dissipated web of feeble, oozing rivulets. These conditions give rise to a meadowy vegetation, whose extending roots still more obstruct the free flow of the waters, and tend to dissipate them out over a yet wider area. Thus the moraine soil and the necessary moisture requisite for the better class of meadow plants are at times combined about as perfectly as if smoothly outspread on a level surface. Where the soil happens to be composed of the finer qualities of glacial detritus and the water is not in excess, the nearest approach is made by the vegetation to that of the lake-meadow. But where, as is more commonly the case, the soil is coarse and bouldery, the vegetation is correspondingly rank. Tall, wide-leaved grasses take their places along the sides, and rushes and nodding carices in the wetter portions, mingled with the most beautiful and imposing flowers,— orange lilies and larkspurs seven or eight feet high, lupines, senecios, aliums, painted-cups, many species of mimulus and pentstemon, the ample boat-leaved *veratrum alba*, and the magnificent alpine columbine, with spurs an inch and a half long. At an elevation of from seven to nine thousand feet showy flowers frequently form the bulk of the vegetation; then the hanging meadows become hanging gardens.

In rare instances we find an alpine basin the bottom of which is a perfect meadow, and the sides nearly all the way round, rising in gentle curves, are covered with moraine soil, which, being saturated with melting snow from encircling fountains, gives rise to an almost continuous girdle of down-curving meadow vegetation that blends gracefully into the level meadow at the bottom, thus forming a grand, smooth, soft, meadow-lined mountain nest. It is in meadows of this sort that the mountain beaver (*Haplodon*) loves to make his home, excavating snug chambers beneath the sod, digging canals, turning the underground waters from channel to channel to suit his convenience, and feeding the vegetation.

Another kind of meadow or bog occurs on densely timbered hillsides where small perennial streams have been dammed at short intervals by fallen trees. Still another kind is found hanging down smooth, flat precipices, while corresponding leaning meadows rise to meet them.

There are also three kinds of small pot-hole meadows one of which is found along the banks of the main streams, another on the summits of rocky ridges, and the third on glacier pavements, all of them interesting in origin and brimful of plant beauty.

CHAPTER VIII

THE FORESTS

THE coniferous forests of the Sierra are the grandest and most beautiful in the world, and grow in a delightful climate on the most interesting and accessible of mountain-ranges, yet strange to say they are not well known. More than sixty years ago David Douglas, an enthusiastic botanist and tree lover, wandered alone through fine sections of the Sugar Pine and Silver Fir woods wild with delight. A few years later, other botanists made short journeys from the coast into the lower woods. Then came the wonderful multitude of miners into the foot-hill zone, mostly blind with gold-dust, soon followed by "sheepmen," who, with wool over their eyes, chased their flocks through all the forest belts from one end of the range to the other. Then the Yosemite Valley was discovered, and thousands of admiring tourists passed through sections of the lower and middle zones on their way to that wonderful park, and gained fine glimpses of the Sugar Pines and Silver Firs along the edges of dusty trails and roads. But few indeed, strong and free with eyes undimmed with care, have gone far enough and lived long enough with the trees to gain anything like a loving conception of their grandeur and significance as manifested in the harmonies of their distribution and varying aspects throughout the seasons, as they stand arrayed in their winter garb rejoicing in storms, putting forth their fresh leaves in the spring while steaming with resiny fragrance, receiving the thunder-showers of summer, or reposing heavy-laden with ripe cones in the rich sungold of autumn. For knowledge of this kind one must dwell with the trees and grow with them, without any reference to time in the almanac sense.

The distribution of the general forest in belts is readily perceived. These, as we have seen, extend in regular order from one extremity of the range to the other; and however dense and somber they may appear in general views, neither on the rocky heights nor down in the leafiest hollows will you find anything to remind you of the dank, malarial selvas of the Amazon and Orinoco, with their "boundless contiguity of shade," the monotonous uniformity of the Deodar forests of the Himalaya, the Black Forest of Europe, or the dense dark woods of Douglas Spruce where rolls the Oregon. The giant pines, and firs, and Sequoias hold their arms open to the sunlight, rising above one another on the mountain benches, marshaled in glorious array, giving forth the utmost expression of grandeur and beauty with inexhaustible variety and harmony.

The inviting openness of the Sierra woods is one of their most distinguishing characteristics. The trees of all the species stand more or less apart in groves, or in small, irregular groups, enabling one to find a way nearly everywhere, along sunny colonnades and through openings that have a smooth,

VIEW IN THE SIERRA FOREST.

park-like surface, strewn with brown needles and burs. Now you cross a wild garden, now a meadow, now a ferny, willowy stream; and ever and anon you emerge from all the groves and flowers upon some granite pavement or high, bare ridge commanding superb views above the waving sea of evergreens far and near.

One would experience but little difficulty in riding on horseback through the successive belts all the way up to the storm-beaten fringes of the icy peaks. The deep cañons, however, that extend from the axis of the range, cut the belts more or less completely into sections, and prevent the mounted traveler from tracing them lengthwise.

This simple arrangement in zones and sections brings the forest, as a whole, within the comprehension of every observer. The different species are ever found occupying the same relative positions to one another, as controlled by soil, climate, and the comparative vigor of each species in taking and holding the ground; and so appreciable are these relations, one need never be at a loss in determining, within a few hundred feet, the elevation above sea-level by the trees alone; for, notwithstanding some of the species range upward for several thousand feet, and all pass one another more or less, yet even those possessing the greatest vertical range are available in this connection, in as much as they take on new forms corresponding with the variations in altitude.

Crossing the treeless plains of the Sacramento and San Joaquin from the west and reaching the Sierra foot-hills, you enter the lower fringe of the forest, composed of small oaks and pines, growing so far apart that not one twentieth of the surface of the ground is in shade at clear noonday. After advancing fifteen or twenty miles, and making an ascent of from two to three thousand feet, you reach

EDGE OF THE TIMBER LINE ON MOUNT SHASTA.

the lower margin of the main pine belt, composed of the gigantic Sugar Pine, Yellow Pine, Incense Cedar, and Sequoia. Next you come to the magnificent Silver Fir belt, and lastly to the upper pine belt, which sweeps up the rocky acclivities of the summit peaks in a dwarfed, wavering fringe to a height of from ten to twelve thousand feet.

This general order of distribution, with reference to climate dependent on elevation, is perceived at once, but there are other harmonies, as far-reaching in this connection, that become manifest only after patient observation and study. Perhaps the most interesting of these is the arrangement of the forests in long, curving bands, braided together into lace-like patterns, and outspread in charming variety. The key to this beautiful harmony is the ancient glaciers; where they flowed the trees followed, tracing their wavering courses along cañons, over ridges, and over high, rolling plateaus. The Cedars of Lebanon, says Hooker, are growing upon one of the moraines of an ancient glacier. All the forests of the Sierra are growing upon moraines. But moraines vanish like the glaciers that make them. Every storm that falls upon them wastes them, cutting gaps, disintegrating boulders, and carrying away their decaying material into new formations, until at length they are no longer recognizable by any save students, who trace their transitional forms down from the fresh moraines still in process of formation, through those that are more and more ancient, and more and more obscured by vegetation and all kinds of post-glacial weathering.

Had the ice-sheet that once covered all the range been melted simultaneously from the foot-hills to the summits, the flanks would, of course, have been left almost bare of soil, and these noble forests would be wanting. Many groves and thickets would undoubtedly have grown up on lake and avalanche beds, and many a fair flower and shrub would have found food and a dwelling-place in weathered nooks

and crevices, but the Sierra as a whole would have
been a bare, rocky desert.

It appears, therefore, that the Sierra forests in
general indicate the extent and positions of the an-

VIEW IN THE MAIN PINE BELT OF THE SIERRA FOREST.

cient moraines as well as they do lines of climate.
For forests, properly speaking, cannot exist without
soil; and, since the moraines have been deposited
upon the solid rock, and only upon elected places,

10

leaving a considerable portion of the old glacial
surface bare, we find luxuriant forests of pine and
fir abruptly terminated by scored and polished
pavements on which not even a moss is growing,
though soil alone is required to fit them for the
growth of trees 200 feet in height.

THE NUT PINE
(*Pinus Sabiniana*)

THE Nut Pine, the first conifer met in ascending the
range from the west, grows only on the torrid foot-
hills, seeming to delight in the most ardent sun-
heat, like a palm; springing up here and there singly,
or in scattered groups of five or six, among scrubby
White Oaks and thickets of ceanothus and manzanita; its extreme upper limit being about 4000 feet
above the sea, its lower about from 500 to 800 feet.

This tree is remarkable for its airy, widespread,
tropical appearance, which suggests a region of
palms, rather than cool, resiny pine woods.   No one
would take it at first sight to be a conifer of any
kind, it is so loose in habit and so widely branched,
and its foliage is so thin and gray.   Full-grown
specimens are from forty to fifty feet in height, and
from two to three feet in diameter.   The trunk
usually divides into three or four main branches,
about fifteen and twenty feet from the ground,
which, after bearing away from one another, shoot
straight up and form separate summits; while the
crooked subordinate branches aspire, and radiate,
and droop in ornamental sprays.   The slender,

grayish-green needles are from eight to twelve inches long, loosely tasseled, and inclined to droop in handsome curves, contrasting with the stiff, dark-

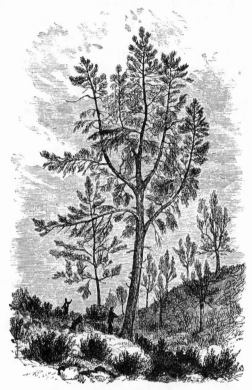

NUT PINE (PINUS SABINIANA).

colored trunk and branches in a very striking manner. No other tree of my acquaintance, so substantial in body, is in its foliage so thin and so pervious to the light. The sunbeams sift through even the leafiest trees with scarcely any interruption, and the weary, heated traveler finds but little protection in their shade.

The generous crop of nutritious nuts which the Nut Pine yields makes it a favorite with Indians, bears, and squirrels. The cones are most beautiful, measuring from five to eight inches in length, and not much less in thickness, rich chocolate-brown in color, and protected by strong, down-curving hooks which terminate the scales. Nevertheless, the little Douglas squirrel can open them. Indians gathering the ripe nuts make a striking picture. The men climb the trees like bears and beat off the cones with sticks, or recklessly cut off the more fruitful branches with hatchets, while the squaws gather the big, generous cones, and roast them until the scales open sufficiently to allow the hard-shelled seeds to be beaten out. Then, in the cool evenings, men, women, and children, with their capacity for dirt greatly increased by the soft resin with which they are all bedraggled, form circles around camp-fires, on the bank of the nearest stream, and lie in easy independence cracking nuts and laughing and chattering, as heedless of the future as the squirrels.

Pinus tuberculata

THIS curious little pine is found at an elevation of from 1500 to 3000 feet, growing in close, willowy groves. It is exceedingly slender and graceful in habit, although trees that chance to stand alone outside the groves sweep forth long, curved branches,

THE GROVE FORM. THE ISOLATED FORM (PINUS TUBERCULATA).

producing a striking contrast to the ordinary grove form. The foliage is of the same peculiar gray-green color as that of the Nut Pine, and is worn about as loosely, so that the body of the tree is scarcely obscured by it.

At the age of seven or eight years it begins to bear cones, not on branches, but on the main axis, and, as they never fall off, the trunk is soon picturesquely dotted with them. The branches also become fruitful after they attain sufficient size. The average size of the older trees is about thirty or forty feet in height, and twelve to fourteen inches in diameter. The cones are about four inches long, exceedingly hard, and covered with a sort of silicious varnish and gum, rendering them impervious to moisture, evidently with a view to the careful preservation of the seeds.

No other conifer in the range is so closely restricted to special localities. It is usually found apart, standing deep in chaparral on sunny hill- and cañon-sides where there is but little depth of soil, and, where found at all, it is quite plentiful; but the ordinary traveler, following carriage-roads and trails, may ascend the range many times without meeting it.

While exploring the lower portion of the Merced Cañon I found a lonely miner seeking his fortune in a quartz vein on a wild mountain-side planted with this singular tree. He told me that he called it the Hickory Pine, because of the whiteness and toughness of the wood. It is so little known, however, that it can hardly be said to have a common name. Most mountaineers refer to it as " that queer little pine-tree covered all over with burs." In my studies of this species I found a very interesting and significant group of facts, whose relations will be seen almost as soon as stated:

1st. All the trees in the groves I examined, however unequal in size, are of the same age.

2d. Those groves are all planted on dry hillsides covered with chaparral, and therefore are liable to be swept by fire.

3d. There are no seedlings or saplings in or about the living groves, but there is always a fine, hopeful crop springing up on the ground once occupied by

LOWER MARGIN OF THE MAIN PINE BELT, SHOWING OPEN CHARACTER OF WOODS.

any grove that has been destroyed by the burning of the chaparral.

4th. The cones never fall off and never discharge their seeds until the tree or branch to which they belong dies.

A full discussion of the bearing of these facts upon one another would perhaps be out of place here, but I may at least call attention to the admirable adaptation of the tree to the fire-swept regions where alone it is found.   After a grove has been destroyed, the ground is at once sown lavishly with all the seeds ripened during its whole life, which seem to have been carefully held in store with reference to such a calamity.   Then a young grove immediately springs up, giving beauty for ashes.

SUGAR PINE
(*Pinus Lambertiana*)

THIS is the noblest pine yet discovered, surpassing all others not merely in size but also in kingly beauty and majesty.

It towers sublimely from every ridge and cañon of the range, at an elevation of from three to seven thousand feet above the sea, attaining most perfect development at a height of about 5000 feet.

Full-grown specimens are commonly about 220 feet high, and from six to eight feet in diameter near the ground, though some grand old patriarch is occasionally met that has enjoyed five or six centuries of storms, and attained a thickness of ten or even twelve feet, living on undecayed, sweet and fresh in every fiber.

In southern Oregon, where it was first discovered by David Douglas, on the head waters of the Umpqua, it attains still grander dimensions, one specimen having been measured that was 245 feet high, and over eighteen feet in diameter three feet from the ground.   The discoverer was the Douglas for whom the noble Douglas Spruce is named, and many other plants which will keep his memory sweet and fresh as long as trees and flowers are loved. His first visit to the Pacific Coast was made in the year 1825. The Oregon Indians watched him with curiosity as he wandered in the woods collecting specimens, and, unlike the fur-gathering strangers they had hitherto known, caring nothing about trade. And when at length they came to know him better, and saw that from year to year the growing things of the woods and prairies were his only objects of pursuit, they called him " The Man of Grass," a title of which he was proud. During his first summer on the waters of the Columbia he made Fort Vancouver his headquarters, making excursions from this Hudson Bay post in every direction. On one of his long trips he saw in an Indian's pouch some of the seeds of a new species of pine which he learned were obtained from a very large tree far to the southward of the Columbia. At the end of the next summer, returning to Fort Vancouver after the setting in of the winter rains, bearing in mind the big pine he had heard of, he set out on an excursion up the Willamette Valley in search of it; and how he fared, and what dangers and hardships he endured, are best told in his own journal, from which I quote as follows:

October 26, 1826. Weather dull. Cold and cloudy. When my friends in England are made acquainted with my travels I fear they will think I have told them nothing but my miseries. . . . I quitted my camp early in the morning to survey the neighboring country, leaving my guide to take charge of the horses until my return in the evening. About an hour's walk from the camp I met an Indian, who on perceiving me instantly strung his bow, placed on his left arm a sleeve of raccoon skin and stood on the defensive. Being quite sure that conduct was prompted by fear and not by hostile intentions, the poor fellow having probably never seen such a being as myself before, I laid my gun at my feet on the ground and waved my hand for him to come to me, which he did slowly and with great caution. I then made him place his bow and quiver of arrows beside my gun, and striking a light gave him a smoke out of my own pipe and a present of a few beads. With my pencil I made a rough sketch of the cone and pine tree which I wanted to obtain, and drew his attention to it, when he instantly pointed with his hand to the hills fifteen or twenty miles distant towards the south; and when I expressed my intention of going thither, cheerfully set out to accompany me. At midday I reached my long-wished-for pines, and lost no time in examining them and endeavoring to collect specimens and seeds. New and strange things seldom fail to make strong impressions, and are therefore frequently over-rated; so that, lest I should never see my friends in England to inform them verbally of this most beautiful and immensely grand tree, I shall here state the dimensions of the largest I could find among several that had been blown down by the wind. At 3 feet from the ground its circumference is 57 feet 9 inches; at 134 feet, 17 feet 5 inches; the extreme length 245 feet. . . . As it was impossible either to climb the tree or hew it down, I endeavored to knock off the cones by firing at them with ball, when the report of my gun brought eight Indians, all of them painted with red earth, armed with bows, arrows, bone-tipped spears, and flint-knives. They appeared anything but friendly. I explained to them what I wanted, and they seemed satisfied and sat down to smoke; but presently I saw one of them string his bow, and another sharpen his flint knife with a pair of wooden pincers and suspend it on the wrist of his right hand. Further testimony of their intentions was unnecessary. To save myself by flight was impossible, so without hesitation I stepped back about five paces, cocked my gun, drew one of the pistols out of my belt, and holding it in my left hand and the gun in my right, showed myself determined to fight for my life. As much as possible I endeavored to preserve my coolness, and thus we stood looking at one another without making any movement or uttering a word for perhaps ten minutes, when one at last, who seemed to be the leader, gave a sign that they wished for some tobacco; this I signified that they should have if they fetched a quantity of cones. They went off immediately in search of them, and no sooner were they all out of sight than I picked up my three cones and some twigs of the trees and made the quickest possible retreat, hurrying back to the camp, which I reached before dusk. . . . I now write lying on the grass with my gun cocked beside me, and penning these lines by the light of my Columbian candle, namely, an ignited piece of rosin-wood.

This grand pine discovered under such exciting circumstances Douglas named in honor of his friend Dr. Lambert of London.

The trunk is a smooth, round, delicately tapered shaft, mostly without limbs, and colored rich purplish-brown, usually enlivened with tufts of yellow lichen. At the top of this magnificent bole, long, curving branches sweep gracefully outward and downward, sometimes forming a palm-like crown, but far more nobly impressive than any palm crown I ever beheld. The needles are about three inches long, finely tempered and arranged in rather close tassels at the ends of slender branchlets that clothe the long, outsweeping limbs. How well they sing in the wind, and how strikingly harmonious an effect

is made by the immense cylindrical cones that depend loosely from the ends of the main branches!
No one knows what Nature can do in the way of
pine-burs until he has seen those of the Sugar
Pine. They are commonly from fifteen to eighteen
inches long, and three in diameter; green, shaded
with dark purple on their sunward sides. They are
ripe in September and October. Then the flat
scales open and the seeds take wing, but the empty
cones become still more beautiful and effective, for
their diameter is nearly doubled by the spreading
of the scales, and their color changes to a warm
yellowish-brown; while they remain swinging on
the tree all the following winter and summer, and
continue effectively beautiful even on the ground
many years after they fall. The wood is deliciously
fragrant, and fine in grain and texture; it is of a rich
cream-yellow, as if formed of condensed sunbeams.
Retinospora obtusa, Siebold, the glory of Eastern
forests, is called " Fu-si-no-ki" (tree of the sun) by
the Japanese; the Sugar Pine is the sun-tree of the
Sierra. Unfortunately it is greatly prized by the
lumbermen, and in accessible places is always
the first tree in the woods to feel their steel. But the
regular lumbermen, with their saw-mills, have been
less generally destructive thus far than the shingle-
makers. The wood splits freely, and there is a constant demand for the shingles. And because an ax,
and saw, and frow are all the capital required for
the business, many of that drifting, unsteady class
of men so large in California engage in it for a few
months in the year. When prospectors, hunters,
ranch hands, etc., touch their " bottom dollar" and

find themselves out of employment, they say, "Well, I can at least go to the Sugar Pines and make shingles." A few posts are set in the ground, and a single length cut from the first tree felled produces boards enough for the walls and roof of a cabin; all the rest the lumberman makes is for sale, and he is speedily independent. No gardener or hay-maker is more sweetly perfumed than these rough mountaineers while engaged in this business, but the havoc they make is most deplorable.

The sugar, from which the common name is derived, is to my taste the best of sweets — better than maple sugar. It exudes from the heart-wood, where wounds have been made, either by forest fires, or the

SUGAR PINE ON EXPOSED RIDGE.

ax, in the shape of irregular, crisp, candy-like kernels, which are crowded together in masses of considerable size, like clusters of resin-beads. When
fresh, it is perfectly white and delicious, but, because most of the wounds on which it is found have
been made by fire, the exuding sap is stained on
the charred surface, and the hardened sugar becomes brown. Indians are fond of it, but on account
of its laxative properties only small quantities may
be eaten. Bears, so fond of sweet things in general, seem never to taste it; at least I have failed to
find any trace of their teeth in this connection.

No lover of trees will ever forget his first meeting
with the Sugar Pine, nor will he afterward need a
poet to call him to "listen what the pine-tree saith."
In most pine-trees there is a sameness of expression,
which, to most people, is apt to become monotonous;
for the typical spiry form, however beautiful, affords
but little scope for appreciable individual character.
The Sugar Pine is as free from conventionalities
of form and motion as any oak. No two are alike,
even to the most inattentive observer; and, notwithstanding they are ever tossing out their immense
arms in what might seem most extravagant gestures,
there is a majesty and repose about them that precludes all possibility of the grotesque, or even picturesque, in their general expression. They are the
priests of pines, and seem ever to be addressing the
surrounding forest. The Yellow Pine is found
growing with them on warm hillsides, and the
White Silver Fir on cool northern slopes; but, noble
as these are, the Sugar Pine is easily king, and
spreads his arms above them in blessing while they

rock and wave in sign of recognition. The main branches are sometimes found to be forty feet in length, yet persistently simple, seldom dividing at all, excepting near the end; but anything like a bare cable appearance is prevented by the small, tasseled branchlets that extend all around them; and when these superb limbs sweep out symmetrically on all sides, a crown sixty or seventy feet wide is formed, which, gracefully poised on the summit of the noble shaft, and filled with sunshine, is one of the most glorious forest objects conceivable. Commonly, however, there is a great preponderance of limbs toward the east, away from the direction of the prevailing winds.

No other pine seems to me so unfamiliar and self-contained. In approaching it, we feel as if in the presence of a superior being, and begin to walk with a light step, holding our breath. Then, perchance, while we gaze awe-stricken, along comes a merry squirrel, chattering and laughing, to break the spell, running up the trunk with no ceremony, and gnawing off the cones as if they were made only for him; while the carpenter-woodpecker hammers away at the bark, drilling holes in which to store his winter supply of acorns.

Although so wild and unconventional when full-grown, the Sugar Pine is a remarkably proper tree in youth. The old is the most original and independent in appearance of all the Sierra evergreens; the young is the most regular,— a strict follower of coniferous fashions,— slim, erect, with leafy, supple branches kept exactly in place, each tapering in outline and terminating in a spiry point. The succes-

sive transitional forms presented between the cautious neatness of youth and bold freedom of maturity offer a delightful study.   At the age of fifty
or sixty years, the shy, fashionable form begins to

YOUNG SUGAR PINE BEGINNING TO BEAR CONES.

be broken up.   Specialized branches push out in
the most unthought-of places, and bend with the
great cones, at once marking individual character,
and this being constantly augmented from year to
year by the varying action of the sunlight, winds,

snow-storms, etc., the individuality of the tree is never again lost in the general forest.

The most constant companion of this species is the Yellow Pine, and a worthy companion it is.

FOREST OF SEQUOIA, SUGAR PINE, AND DOUGLAS SPRUCE.

The Douglas Spruce, Libocedrus, Sequoia, and the White Silver Fir are also more or less associated with it; but on many deep-soiled mountain-sides,

11

at an elevation of about 5000 feet above the sea, it forms the bulk of the forest, filling every swell and hollow and down-plunging ravine. The majestic crowns, approaching each other in bold curves, make a glorious canopy through which the tempered sunbeams pour, silvering the needles, and gilding the massive boles, and flowery, park-like ground, into a scene of enchantment.

On the most sunny slopes the white-flowered fragrant chamoebatia is spread like a carpet, brightened during early summer with the crimson Sarcodes, the wild rose, and innumerable violets and gilias. Not even in the shadiest nooks will you find any rank, untidy weeds or unwholesome darkness. On the north sides of ridges the boles are more slender, and the ground is mostly occupied by an underbrush of hazel, ceanothus, and flowering dogwood, but never so densely as to prevent the traveler from sauntering where he will; while the crowning branches are never impenetrable to the rays of the sun, and never so interblended as to lose their individuality.

View the forest from beneath or from some commanding ridge-top; each tree presents a study in itself, and proclaims the surpassing grandeur of the species.

YELLOW, OR SILVER PINE
(*Pinus ponderosa*)

THE Silver, or Yellow, Pine, as it is commonly called, ranks second among the pines of the Sierra as a lumber tree, and almost rivals the Sugar Pine in stature and nobleness of port. Because of its superior powers of enduring variations of climate and soil, it has a more extensive range than any other conifer growing on the Sierra. On the western slope it is first met at an elevation of about 2000 feet, and extends nearly to the upper limit of the timber line. Thence, crossing the range by the lowest passes, it descends to the eastern base, and pushes out for a considerable distance into the hot volcanic plains, growing bravely upon well-watered moraines, gravelly lake basins, arctic ridges, and torrid lava-beds; planting itself upon the lips of craters, flourishing vigorously even there, and tossing ripe cones among the ashes and cinders of Nature's hearths.

The average size of full-grown trees on the western slope, where it is associated with the Sugar Pine, is a little less than 200 feet in height and from five to six feet in diameter, though specimens may easily be found that are considerably larger. I measured one, growing at an elevation of 4000 feet in the valley of the Merced, that is a few inches over eight feet in diameter, and 220 feet high.

Where there is plenty of free sunshine and other conditions are favorable, it presents a striking contrast in form to the Sugar Pine, being a symmetrical spire, formed of a straight round trunk, clad with innumerable branches that are divided over and over again. About one half of the trunk is commonly branchless, but where it grows at all close, three fourths or more become naked; the tree presenting then a more slender and elegant shaft than any other tree in the woods. The bark is mostly arranged in massive plates, some of them measuring

PINUS PONDEROSA.

four or five feet in length by eighteen inches in
width, with a thickness of three or four inches,

forming a quite marked and distinguishing feature. The needles are of a fine, warm, yellow-green color, six to eight inches long, firm and elastic, and crowded in handsome, radiant tassels on the upturning ends of the branches. The cones are about three or four inches long, and two and a half wide, growing in close, sessile clusters among the leaves.

The species attains its noblest form in filled-up lake basins, especially in those of the older yosemites, and so prominent a part does it form of their groves that it may well be called the Yosemite Pine. Ripe specimens favorably situated are almost always 200 feet or more in height, and the branches clothe the trunk nearly to the ground, as seen in the illustration.

The Jeffrey variety attains its finest development in the northern portion of the range, in the wide basins of the McCloud and Pitt rivers, where it forms magnificent forests scarcely invaded by any other tree. It differs from the ordinary form in size, being only about half as tall, and in its redder and more closely furrowed bark, grayish-green foliage, less divided branches, and larger cones; but intermediate forms come in which make a clear separation impossible, although some botanists regard it as a distinct species. It is this variety that climbs storm-swept ridges, and wanders out among the volcanoes of the Great Basin. Whether exposed to extremes of heat or cold, it is dwarfed like every other tree, and becomes all knots and angles, wholly unlike the majestic forms we have been sketching. Old specimens, bearing cones about as big as pineapples, may sometimes be found clinging to rifted

SILVER PINE 210 FEET HIGH.
(THE FORM GROWING IN
YOSEMITE VALLEY.)

rocks at an elevation of seven or eight thousand feet, whose highest branches scarce reach above one's shoulders.

I have oftentimes feasted on the beauty of these noble trees when they were towering in all their winter grandeur, laden with snow — one mass of bloom; in summer, too, when the brown, staminate clusters hang thick among the shimmering needles, and the big purple burs are ripening in the mellow light; but it is during cloudless wind-storms that these colossal pines are most impressively beautiful. Then they bow like willows, their leaves streaming forward all in one direction, and, when the sun shines upon them at the required angle, entire groves glow as if every leaf were burnished silver. The fall of tropic light on the royal crown of a palm is a truly glorious spectacle, the fervid sun-flood breaking upon the glossy leaves in long lance-rays, like mountain water among boulders. But to me there is something more impressive in the fall of light upon these Silver Pines. It seems beaten to the finest dust, and is shed off in myriads of minute sparkles that seem to come from the very heart of the trees, as if, like rain falling upon fertile soil, it had been absorbed, to reappear in flowers of light.

This species also gives forth the finest music to the wind. After listening to it in all kinds of winds, night and day, season after season, I think I could approximate to my position on the mountains by this pine-music alone. If you would catch the tones of separate needles, climb a tree. They are well tempered, and give forth no uncertain sound, each standing out, with no interference excepting during heavy gales; then you may detect the click of one·needle upon another, readily distinguishable from their free, wing-like hum. Some idea of their temper may be drawn from the fact that, notwithstanding they are so long, the vibrations that give rise to the peculiar shimmering of the light are made at the rate of about two hundred and fifty per minute.

When a Sugar Pine and one of this species equal in size are observed together, the latter is seen to be far more simple in manners, more lithely graceful, and its beauty is of a kind more easily appreciated; but then, it is, on the other hand, much less dignified and original in demeanor. The Silver Pine seems eager to shoot aloft. Even while it is drowsing in autumn sun-gold, you may still detect a skyward aspiration. But the Sugar Pine seems too unconsciously noble, and too complete in every way, to leave room for even a heavenward care.

DOUGLAS SPRUCE

(*Pseudotsuga Douglasii*)

THIS tree is the king of the spruces, as the Sugar Pine is king of pines. It is by far the most majestic spruce I ever beheld in any forest, and one of the largest and longest lived of the giants that flourish throughout the main pine belt, often attaining a height of nearly 200 feet, and a diameter of six or seven. Where the growth is not too close, the strong, spreading branches come more than halfway down the trunk, and these are hung with innumerable slender, swaying sprays, that are handsomely feathered with the short leaves which radiate at right angles all around them. This vigorous spruce is ever beautiful, welcoming the mountain winds and the snow as well as the mellow summer light, and maintaining its youthful freshness undiminished from century to century through a thousand storms.

It makes its finest appearance in the months of June and July. The rich brown buds with which its sprays are tipped swell and break about this time, revealing the young leaves, which at first are bright yellow, making the tree appear as if covered with gay blossoms; while the pendulous bracted cones with their shell-like scales are a constant adornment.

The young trees are mostly gathered into beautiful family groups, each sapling exquisitely symmetrical. The primary branches are whorled regularly around the axis, generally in fives, while each is draped with long, feathery sprays, that descend in curves as free and as finely drawn as those of falling water.

In Oregon and Washington it grows in dense forests, growing tall and mast-like to a height of 300 feet, and is greatly prized as a lumber tree. But in the Sierra it is scattered among other trees, or forms small groves, seldom ascending higher than 5500 feet, and never making what would be called a forest. It is not particular in its choice of soil — wet or dry, smooth or rocky, it makes out to live well on them all. Two of the largest specimens I have measured are in Yosemite Valley, one of which is more than eight feet in diameter, and is growing upon the terminal moraine of the residual glacier that occupied the South Fork Cañon; the other is nearly as large, growing upon angular blocks of granite that have been shaken from the precipitous front of the Liberty Cap near the Nevada Fall. No other tree seems so capable of adapting itself to earthquake taluses, and many of these rough boulder-slopes are occupied by it almost exclusively, especially in yosemite gorges moistened by the spray of waterfalls.

INCENSE CEDAR
(*Libocedrus decurrens*)

THE Incense Cedar is another of the giants quite generally distributed throughout this portion of the forest, without exclusively occupying any considerable area, or even making extensive groves. It ascends to about 5000 feet on the warmer hillsides, and reaches the climate most congenial to it at about from 3000 to 4000 feet, growing vigorously at this elevation on all kinds of soil, and in particular it is capable of enduring more moisture about its roots than any of its companions, excepting only the Sequoia.

The largest specimens are about 150 feet high, and seven feet in diameter. The bark is brown, of a singularly rich tone very attractive to artists, and the foliage is tinted with a warmer yellow than that of any other evergreen in the woods. Casting your eye over the general forest from some ridge-top, the color alone of its spiry summits is sufficient to identify it in any company.

In youth, say up to the age of seventy or eighty years, no other tree forms so strictly tapered a cone from top to bottom. The branches swoop outward and downward in bold curves, excepting the younger ones near the top, which aspire, while the lowest droop to the ground, and all spread out in flat, ferny plumes, beautifully fronded, and imbricated upon one another. As it becomes older, it grows strikingly irregular and picturesque. Large special branches put out at right angles from the trunk, form big, stubborn elbows, and then shoot up parallel with the axis. Very old trees are usually dead at the top, the main axis protruding above ample masses of green plumes, gray and lichen-covered, and drilled full of acorn holes by the woodpeckers. The plumes are exceedingly beautiful; no waving fern-frond in shady dell is more unreservedly beautiful in form and texture, or half so inspiring in color and spicy fragrance. In its prime, the whole tree is thatched with them, so that they shed off rain and snow like a roof, making fine mansions for storm-bound birds and mountaineers. But if you would see the *Libocedrus* in all its glory, you must

INCENSE CEDAR IN ITS PRIME.

go to the woods in winter. Then it is laden with myriads of four-sided staminate cones about the size of wheat grains,— winter wheat,— producing a golden tinge, and forming a noble illustration of Nature's immortal vigor and virility. The fertile cones are about three fourths of an inch long, borne on the outside of the plumy branchlets, where they serve to enrich still more the surpassing beauty of this grand winter-blooming goldenrod.

WHITE SILVER FIR
(*Abies concolor*)

We come now to the most regularly planted of

FOREST OF GRAND SILVER FIRS. TWO SEQUOIAS IN THE FOREGROUND ON THE LEFT.

all the main forest belts, composed almost exclusively of two noble firs—*A. concolor* and *A. magnifica*. It extends with no marked interruption for 450 miles, at an elevation of from 5000 to nearly 9000 feet above the sea. In its youth *A. concolor* is a charmingly symmetrical tree with branches regularly whorled in level collars around its whitish-gray axis, which terminates in a strong,

hopeful shoot. The leaves are in two horizontal
rows, along branchlets that commonly are less than
eight years old, forming handsome plumes, pin-
nated like the fronds of ferns. The cones are grayish-green when ripe, cylindrical, about from three
to four inches long by one and a half to two inches
wide, and stand upright on the upper branches.

Full-grown trees, favorably situated as to soil
and exposure, are about 200 feet high, and five or
six feet in diameter near the ground, though larger
specimens are by no means rare.

As old age creeps on, the bark becomes rougher
and grayer, the branches lose their exact regularity,
many are snow-bent or broken off, and the main
axis often becomes double or otherwise irregular
from accidents to the terminal bud or shoot; but
throughout all the vicissitudes of its life on the
mountains, come what may, the noble grandeur of
the species is patent to every eye.

MAGNIFICENT SILVER FIR, OR RED FIR

(Abies magnifica)

THIS is the most charmingly symmetrical of all
the giants of the Sierra woods, far surpassing its
companion species in this respect, and easily distinguished from it by the purplish-red bark, which
is also more closely furrowed than that of the white,
and by its larger cones, more regularly whorled and
fronded branches, and by its leaves, which are
shorter, and grow all around the branchlets and
point upward.

In size, these two Silver Firs are about equal, the *magnifica* perhaps a little the taller. Specimens from 200 to 250 feet high are not rare on well-ground moraine soil, at an elevation of from 7500 to 8500 feet above sea-level. The largest that I measured stands back three miles from the brink of the north wall of Yosemite Valley. Fifteen years ago it was 240 feet high, with a diameter of a little more than five feet.

Happy the man with the freedom and the love to climb one of these superb trees in full flower and fruit. How admirable the forest-work of Nature is then seen to be, as one makes his way up through the midst of the broad, fronded branches, all arranged in exquisite order around the trunk, like the whorled leaves of lilies, and each branch and branchlet about as strictly pinnate as the most symmetrical fern-frond. The staminate cones are seen growing straight downward from the under side of the young branches in lavish profusion, making fine purple clusters amid the grayish-green foliage. On the topmost branches the fertile cones are set firmly on end like small casks. They are about six inches long, three wide, covered with a fine gray down, and streaked with crystal balsam that seems to have been poured upon each cone from above.

Both the Silver Firs live 250 years or more when the conditions about them are at all favorable. Some venerable patriarch may often be seen, heavily storm-marked, towering in severe majesty above the rising generation, with a protecting grove of saplings pressing close around his feet, each dressed with such loving care that not a leaf seems want-

VIEW OF FOREST OF THE MAGNIFICENT SILVER FIR.

ing. Other companies are made up of trees near
the prime of life, exquisitely harmonized to one
another in form and gesture, as if Nature had culled
them one by one with nice discrimination from all
the rest of the woods.

It is from this tree, called Red Fir by the lumberman, that mountaineers always cut boughs to sleep
on when they are so fortunate as to be within its
limits. Two rows of the plushy branches overlapping along the middle, and a crescent of smaller
plumes mixed with ferns and flowers for a pillow,
form the very best bed imaginable. The essences
of the pressed leaves seem to fill every pore of one's
body, the sounds of falling water make a soothing
hush, while the spaces between the grand spires
afford noble openings through which to gaze
dreamily into the starry sky. Even in the matter
of sensuous ease, any combination of cloth, steel
springs, and feathers seems vulgar in comparison.

The fir woods are delightful sauntering-grounds
at any time of year, but most so in autumn. Then
the noble trees are hushed in the hazy light, and
drip with balsam; the cones are ripe, and the seeds,
with their ample purple wings, mottle the air like
flocks of butterflies; while deer feeding in the
flowery openings between the groves, and birds and
squirrels in the branches, make a pleasant stir which
enriches the deep, brooding calm of the wilderness,
and gives a peculiar impressiveness to every tree.
No wonder the enthusiastic Douglas went wild with
joy when he first discovered this species. Even in
the Sierra, where so many noble evergreens chal-
lenge admiration, we linger among these colossal firs

SILVER-FIR FOREST GROWING ON MORAINES OF THE HOFFMAN AND TENAYA GLACIERS.

with fresh love, and extol their beauty again and again, as if no other in the world could henceforth claim our regard.

It is in these woods the great granite domes rise that are so striking and characteristic a feature of the Sierra. And here too we find the best of the garden meadows. They lie level on the tops of the dividing ridges, or sloping on the sides of them, embedded in the magnificent forest. Some of these meadows are in great part occupied by *Veratrum alba*, which here grows rank and tall, with boat-shaped leaves thirteen inches long and twelve inches wide, ribbed like those of cypripedium. Columbine grows on the drier margins with tall larkspurs and lupines waist-deep in grasses and sedges; several species of castilleia also make a bright show in beds of blue and white violets and daisies. But the glory of these forest meadows is a lily — *L. parvum.* The flowers are orange-colored and quite small, the smallest I ever saw of the true lilies; but it is showy nevertheless, for it is seven to eight feet high and waves magnificent racemes of ten to twenty flowers or more over one's head, while it stands out in the open ground with just enough of grass and other plants about it to make a fringe for its feet and show it off to best advantage.

A dry spot a little way back from the margin of a Silver Fir lily garden makes a glorious camp-ground, especially where the slope is toward the east and opens a view of the distant peaks along the summit of the range. The tall lilies are brought forward in all their glory by the light of your blazing camp-fire, relieved against the outer darkness, and the nearest of the trees with their whorled branches tower above you like larger lilies, and the sky seen through the garden opening seems one vast meadow of white lily stars.

In the morning everything is joyous and bright, the delicious purple of the dawn changes softly to daffodil yellow and white; while the sunbeams pouring through the passes between the peaks give a margin of gold to each of them. Then the spires of the firs in the hollows of the middle region catch the glow, and your camp grove is filled with light. The birds begin to stir, seeking sunny branches on the edge of the meadow for sun-baths after the cold night, and looking for their breakfasts, every one of them as fresh as a lily and as charmingly arrayed. Innumerable insects begin to dance, the deer withdraw from the open glades and ridge-tops to their leafy hiding-places in the chaparral, the flowers open and straighten their petals as the dew vanishes, every pulse beats high, every life-cell rejoices, the very rocks seem to tingle with life, and God is felt brooding over everything great and small.

BIG TREE

(Sequoia gigantea)

BETWEEN the heavy pine and Silver Fir belts we find the Big Tree, the king of all the conifers in the world, " the noblest of a noble race." It extends in a widely interrupted belt from a small grove on the middle fork of the American River to the head of Deer Creek, a distance of about 260 miles, the northern limit being near the thirty-ninth parallel, the southern a little below the thirty-sixth, and the elevation of the belt above the sea varies from about 5000 to 8000 feet. From the American River grove to the forest on King's River the species occurs only in small isolated groups so sparsely distributed along the belt that three of the gaps in it are from forty to sixty miles wide. But from King's River southward the Sequoia is not restricted to mere groves, but extends across the broad rugged basins of the Kaweah and Tule rivers in noble forests, a distance of nearly seventy miles, the continuity of this part of the belt being broken only by deep cañons. The Fresno, the largest of the northern groves, occupies an area of three or four square miles, a short distance to the southward of the famous Mariposa Grove. Along the beveled rim of the cañon of the south fork of King's River there is a majestic forest of Sequoia about six miles long by two wide. This is the northernmost assemblage of Big Trees that may fairly be called a forest. Descending the precipitous divide between the King's River and Kaweah you enter the grand forests that form the main continuous portion of the belt. Advancing southward the giants become more and more irrepressibly exuberant, heaving their massive crowns into the sky from every ridge and slope, and waving onward in graceful compliance with the complicated topography of the region. The finest of the Kaweah section of the belt is on the broad ridge between Marble Creek and the middle fork, and extends from the granite headlands overlooking the hot plains to within a few

SEQUOIA GIGANTEA—VIEW IN GENERAL GRANT NATIONAL PARK.

miles of the cool glacial fountains of the summit peaks. The extreme upper limit of the belt is reached between the middle and south forks of the Kaweah at an elevation of 8400 feet. But the finest block of Big Tree forest in the entire belt is on the north fork of Tule River. In the northern groves there are comparatively few young trees or saplings. But here for every old, storm-stricken giant there are many in all the glory of prime vigor, and for each of these a crowd of eager, hopeful young trees and saplings growing heartily on moraines, rocky ledges, along watercourses, and in the moist alluvium of meadows, seemingly in hot pursuit of eternal life.

But though the area occupied by the species increases so much from north to south there is no marked increase in the size of the trees. A height of 275 feet and a diameter near the ground of about 20 feet is perhaps about the average size of full-grown trees favorably situated; specimens 25 feet in diameter are not very rare, and a few are nearly 300 feet high. In the Calaveras Grove there are four trees over 300 feet in height, the tallest of which by careful measurement is 325 feet. The largest I have yet met in the course of my explorations is a majestic old scarred monument in the King's River forest. It is 35 feet 8 inches in diameter inside the bark four feet from the ground. Under the most favorable conditions these giants probably live 5000 years or more, though few of even the larger trees are more than half as old. I never saw a Big Tree that had died a natural death; barring accidents they seem to be immortal, being exempt from all the diseases that afflict and kill other trees. Unless destroyed by man, they live on indefinitely until burned, smashed by lightning, or cast down by storms, or by the giving way of the ground on which they stand. The age of one that was felled in the Calaveras Grove, for the sake of having its stump for a dancing-floor, was about 1300 years, and its diameter, measured across the stump, 24 feet inside the bark. Another that was cut down in the King's River forest was about the same size, but nearly a thousand years older (2200 years), though not a very old-looking tree. It was felled to procure a section for exhibition, and thus an opportunity was given to count its annual rings of growth. The colossal scarred monument in the King's River forest mentioned above is burned half through, and I spent a day in making an estimate of its age, clearing away the charred surface with an ax and carefully counting the annual rings with the aid of a pocket-lens. The wood-rings in the section I laid bare were so involved and contorted in some places that I was not able to determine its age exactly, but I counted over 4000 rings, which showed that this tree was in its prime, swaying in the Sierra winds, when Christ walked the earth. No other tree in the world, as far as I know, has looked down on so many centuries as the Sequoia, or opens such impressive and suggestive views into history.

So exquisitely harmonious and finely balanced are even the very mightiest of these monarchs of the woods in all their proportions and circumstances there never is anything overgrown or monstrous-

looking about them. On coming in sight of them
for the first time, you are likely to say, "Oh, see
what beautiful, noble-looking trees are towering
there among the firs and pines!"—their grandeur
being in the mean time in great part invisible, but
to the living eye it will be manifested sooner or
later, stealing slowly on the senses, like the grandeur of Niagara, or the lofty Yosemite domes. Their
great size is hidden from the inexperienced observer
as long as they are seen at a distance in one harmonious view. When, however, you approach them
and walk round them, you begin to wonder at their
colossal size and seek a measuring-rod. These
giants bulge considerably at the base, but not more
than is required for beauty and safety; and the
only reason that this bulging seems in some cases
excessive is that only a comparatively small section
of the shaft is seen at once in near views. One
that I measured in the King's River forest was 25
feet in diameter at the ground, and 10 feet in
diameter 200 feet above the ground, showing that
the taper of the trunk as a whole is charmingly
fine. And when you stand back far enough to see
the massive columns from the swelling instep to
the lofty summit dissolving in a dome of verdure,
you rejoice in the unrivaled display of combined
grandeur and beauty. About a hundred feet or
more of the trunk is usually branchless, but its massive simplicity is relieved by the bark furrows,
which instead of making an irregular network run
evenly parallel, like the fluting of an architectural
column, and to some extent by tufts of slender
sprays that wave lightly in the winds and cast

flecks of shade, seeming to have been pinned on here and there for the sake of beauty only. The young trees have slender simple branches down to the ground, put on with strict regularity, sharply aspiring at the top, horizontal about half-way down, and drooping in handsome curves at the base. By the time the sapling is five or six hundred years old this spiry, feathery, juvenile habit merges into the firm, rounded dome form of middle age, which in turn takes on the eccentric picturesqueness of old age. No other tree in the Sierra forest has foliage so densely massed or presents outlines so firmly drawn and so steadily subordinate to a special type. A knotty ungovernable-looking branch five to eight feet thick may be seen pushing out abruptly from the smooth trunk, as if sure to throw the regular curve into confusion, but as soon as the general outline is reached it stops short and dissolves in spreading bosses of law-abiding sprays, just as if every tree were growing beneath some huge, invisible bell-glass, against whose sides every branch was being pressed and molded, yet somehow indulging in so many small departures from the regular form that there is still an appearance of freedom.

The foliage of the saplings is dark bluish-green in color, while the older trees ripen to a warm brownish-yellow tint like Libocedrus. The bark is rich cinnamon-brown, purplish in young trees and in shady portions of the old, while the ground is covered with brown leaves and burs forming color-masses of extraordinary richness, not to mention the flowers and underbrush that rejoice about them in their seasons. Walk the Sequoia woods at any time

of year and you will say they are the most beautiful and majestic on earth. Beautiful and impressive
contrasts meet you everywhere : the colors of tree
and flower, rock and sky, light and shade, strength
and frailty, endurance and evanescence, tangles of
supple hazel-bushes, tree-pillars about as rigid as
granite domes, roses and violets, the smallest of their
kind, blooming around the feet of the giants, and
rugs of the lowly chamæbatia where the sunbeams
fall. Then in winter the trees themselves break
forth in bloom, myriads of small four-sided staminate cones crowd the ends of the slender sprays,
coloring the whole tree, and when ripe dusting the
air and the ground with golden pollen. The fertile
cones are bright grass-green, measuring about two
inches in length by one and a half in thickness,
and are made up of about forty firm rhomboidal
scales densely packed, with from five to eight seeds
at the base of each. A single cone, therefore, contains from two to three hundred seeds, which are
about a fourth of an inch long by three sixteenths
wide, including a thin, flat margin that makes them
go glancing and wavering in their fall like a boy's
kite. The fruitfulness of Sequoia may be illustrated
by two specimen branches one and a half and two
inches in diameter on which I counted 480 cones.
No other Sierra conifer produces nearly so many
seeds. Millions are ripened annually by a single
tree, and in a fruitful year the product of one of the
northern groves would be enough to plant all the
mountain-ranges of the world. Nature takes care,
however, that not one seed in a million shall germinate at all, and of those that do perhaps not one

in ten thousand is suffered to live through the many
vicissitudes of storm, drought, fire, and snow-crush-
ing that beset their youth.

The Douglas squirrel is the happy harvester of
most of the Sequoia cones.   Out of every hundred
perhaps ninety fall to his share, and unless cut off
by his ivory sickle they shake out their seeds and
remain on the tree for many years.   Watching the
squirrels at their harvest work in the Indian summer is one of the most delightful diversions imaginable.   The woods are calm and the ripe colors are
blazing in all their glory; the cone-laden trees stand
motionless in the warm, hazy air, and you may see
the crimson-crested woodcock, the prince of Sierra
woodpeckers, drilling some dead limb or fallen trunk
with his bill, and ever and anon filling the glens
with his happy cackle.   The humming-bird, too,
dwells in these noble woods, and may oftentimes be
seen glancing among the flowers or resting wing-
weary on some leafless twig; here also are the familiar robin of the orchards, and the brown and
grizzly bears so obviously fitted for these majestic
solitudes; and the Douglas squirrel, making more
hilarious, exuberant, vital stir than all the bears,
birds, and humming wings together.

As soon as any accident happens to the crown of
these Sequoias, such as being stricken off by lightning or broken by storms, then the branches beneath the wound, no matter how situated, seem to
be excited like a colony of bees that have lost their
queen, and become anxious to repair the damage.
Limbs that have grown outward for centuries at
right angles to the trunk begin to turn upward to

assist in making a new crown, each speedily assuming the special form of true summits. Even in the case of mere stumps, burned half through, some mere ornamental tuft will try to go aloft and do its best as a leader in forming a new head.

Groups of two or three of these grand trees are often found standing close together, the seeds from which they sprang having probably grown on ground cleared for their reception by the fall of a large tree of a former generation. These patches of fresh, mellow soil beside the upturned roots of the fallen giant may be from forty to sixty feet wide, and they are speedily occupied by seedlings. Out of these seedling-thickets perhaps two or three may become trees, forming those close groups called "three graces," "loving couples," etc. For even supposing that the trees should stand twenty or thirty feet apart while young, by the time they are full-grown their trunks will touch and crowd against each other and even appear as one in some cases.

It is generally believed that this grand Sequoia was once far more widely distributed over the Sierra; but after long and careful study I have come to the conclusion that it never was, at least since the close of the glacial period, because a diligent search along the margins of the groves, and in the gaps between, fails to reveal a single trace of its previous existence beyond its present bounds. Notwithstanding, I feel confident that if every Sequoia in the range were to die to-day, numerous monuments of their existence would remain, of so imperishable a nature as to be available for the student more than ten thousand years hence.

In the first place we might notice that no species of coniferous tree in the range keeps its individuals so well together as Sequoia; a mile is perhaps the greatest distance of any straggler from the main body, and all of those stragglers that have come under my observation are young, instead of old monumental trees, relics of a more extended growth.

Again, Sequoia trunks frequently endure for centuries after they fall. I have a specimen block, cut from a fallen trunk, which is hardly distinguishable from specimens cut from living trees, although the old trunk-fragment from which it was derived has lain in the damp forest more than 380 years, probably thrice as long. The time measure in the case is simply this: when the ponderous trunk to which the old vestige belonged fell, it sunk itself into the ground, thus making a long, straight ditch, and in the middle of this ditch a Silver Fir is growing that is now four feet in diameter and 380 years old, as determined by cutting it half through and counting the rings, thus demonstrating that the remnant of the trunk that made the ditch has lain on the ground *more* than 380 years. For it is evident that to find the whole time, we must add to the 380 years the time that the vanished portion of the trunk lay in the ditch before being burned out of the way, plus the time that passed before the seed from which the monumental fir sprang fell into the prepared soil and took root. Now, because Sequoia trunks are never wholly consumed in one forest fire, and those fires recur only at considerable intervals, and because Sequoia ditches after being cleared are often left unplanted for centuries, it becomes evident that the trunk remnant in question may probably have lain a thousand years or more. And this instance is by no means a rare one.

But admitting that upon those areas supposed to have been once covered with Sequoia every tree may have fallen, and every trunk may have been burned or buried, leaving not a remnant, many of the ditches made by the fall of the ponderous trunks, and the bowls made by their upturning roots, would remain patent for thousands of years after the last vestige of the trunks that made them had vanished. Much of this ditch-writing would no doubt be quickly effaced by the flood-action of overflowing streams and rain-washing; but no inconsiderable portion would remain enduringly engraved on ridge-tops beyond such destructive action; for, where all the conditions are favorable, it is almost imperishable. *Now these historic ditches and root bowls occur in all the present Sequoia groves and forests, but as far as I have observed, not the faintest vestige of one presents itself outside of them.*

We therefore conclude that the area covered by Sequoia has not been diminished during the last eight or ten thousand years, and probably not at all in post-glacial times.

Is the species verging to extinction? What are its relations to climate, soil, and associated trees?

All the phenomena bearing on these questions also throw light, as we shall endeavor to show, upon the peculiar distribution of the species, and sustain the conclusion already arrived at on the question of extension.

In the northern groups, as we have seen, there

are few young trees or saplings growing up around
the failing old ones to perpetuate the race, and in
as much as those aged Sequoias, so nearly child-
less, are the only ones commonly known, the species,
to most observers, seems doomed to speedy extinction, as being nothing more than an expiring remnant, vanquished in the so-called struggle for life
by pines and firs that have driven it into its last
strongholds in moist glens where climate is exceptionally favorable. But the language of the majestic continuous forests of the south creates a very
different impression. No tree of all the forest is
more enduringly established in concordance with
climate and soil. It grows heartily everywhere —
on moraines, rocky ledges, along watercourses, and
in the deep, moist alluvium of meadows, with a multitude of seedlings and saplings crowding up around
the aged, seemingly abundantly able to maintain
the forest in prime vigor. For every old storm-
stricken tree, there is one or more in all the glory of
prime; and for each of these many young trees and
crowds of exuberant saplings. So that if all the
trees of any section of the main Sequoia forest were
ranged together according to age, a very promising
curve would be presented, all the way up from last
year's seedlings to giants, and with the young and
middle-aged portion of the curve many times longer
than the old portion. Even as far north as the
Fresno, I counted 536 saplings and seedlings growing promisingly upon a piece of rough avalanche
soil not exceeding two acres in area. This soil bed
is about seven years old, and has been seeded almost simultaneously by pines, firs, Libocedrus, and

MUIR GORGE, TUOLUMNE CAÑON—YOSEMITE NATIONAL PARK.

Sequoia, presenting a simple and instructive illustration of the struggle for life among the rival species; and it was interesting to note that the conditions thus far affecting them have enabled the young Sequoias to gain a marked advantage.

In every instance like the above I have observed that the seedling Sequoia is capable of growing on both drier and wetter soil than its rivals, but requires more sunshine than they; the latter fact being clearly shown wherever a Sugar Pine or fir is growing in close contact with a Sequoia of about equal age and size, and equally exposed to the sun; the branches of the latter in such cases are always less leafy. Toward the south, however, where the Sequoia becomes *more* exuberant and numerous, the rival trees become *less* so; and where they mix with Sequoias, they mostly grow up beneath them, like slender grasses among stalks of Indian corn. Upon a bed of sandy flood-soil I counted ninety-four Sequoias, from one to twelve feet high, on a patch of ground once occupied by four large Sugar Pines which lay crumbling beneath them,— an instance of conditions which have enabled Sequoias to crowd out the pines.

I also noted eighty-six vigorous saplings upon a piece of fresh ground prepared for their reception by fire. Thus fire, the great destroyer of Sequoia, also furnishes bare virgin ground, one of the conditions essential for its growth from the seed. Fresh ground is, however, furnished in sufficient quantities for the constant renewal of the forests without fire, viz., by the fall of old trees. The soil is thus upturned and mellowed, and many trees are planted for every one that falls. Land-slips and floods also give rise to bare virgin ground; and a tree now and then owes its existence to a burrowing wolf or squirrel, but the most regular supply of fresh soil is furnished by the fall of aged trees.

The climatic changes in progress in the Sierra, bearing on the tenure of tree life, are entirely misapprehended, especially as to the time and the means employed by Nature in effecting them. It is constantly asserted in a vague way that the Sierra was vastly wetter than now, and that the increasing drought will of itself extinguish Sequoia, leaving its ground to other trees supposed capable of flourishing in a drier climate. But that Sequoia can and does grow on as dry ground as any of its present rivals, is manifest in a thousand places. "Why, then," it will be asked, "are Sequoias always found in greatest abundance in well-watered places where streams are exceptionally abundant?" Simply because a growth of Sequoias creates those streams. The thirsty mountaineer knows well that in every Sequoia grove he will find running water, but it is a mistake to suppose that the water is the cause of the grove being there; on the contrary, the grove is the cause of the water being there. Drain off the water and the trees will remain, but cut off the trees, and the streams will vanish. Never was cause more completely mistaken for effect than in the case of these related phenomena of Sequoia woods and perennial streams, and I confess that at first I shared in the blunder.

When attention is called to the method of Sequoia stream-making, it will be apprehended at once.

The roots of this immense tree fill the ground, forming a thick sponge that absorbs and holds back the rains and melting snows, only allowing them to ooze and flow gently. Indeed, every fallen leaf and rootlet, as well as long clasping root, and prostrate trunk, may be regarded as a dam hoarding the bounty of storm-clouds, and dispensing it as blessings all through the summer, instead of allowing it to go headlong in short-lived floods. Evaporation is also checked by the dense foliage to a greater extent than by any other Sierra tree, and the air is entangled in masses and broad sheets that are quickly saturated; while thirsty winds are not allowed to go sponging and licking along the ground.

So great is the retention of water in many places in the main belt, that bogs and meadows are created by the killing of the trees. A single trunk falling across a stream in the woods forms a dam 200 feet long, and from ten to thirty feet high, giving rise to a pond which kills the trees within its reach. These dead trees fall in turn, thus making a clearing, while sediments gradually accumulate changing the pond into a bog, or meadow, for a growth of carices and sphagnum. In some instances a series of small bogs or meadows rise above one another on a hillside, which are gradually merged into one another, forming sloping bogs, or meadows, which make striking features of Sequoia woods, and since all the trees that have fallen into them have been preserved, they contain records of the generations that have passed since they began to form.

Since, then, it is a fact that thousands of Sequoias are growing thriftily on what is termed dry ground,

13

and even clinging like mountain pines to rifts in granite precipices; and since it has also been shown that the extra moisture found in connection with the denser growths is an effect of their presence, instead of a cause of their presence, then the notions as to the former extension of the species and its near approach to extinction, based upon its supposed dependence on greater moisture, are seen to be erroneous.

The decrease in the rain- and snowfall since the close of the glacial period in the Sierra is much less than is commonly guessed. The highest post-glacial watermarks are well preserved in all the upper river channels, and they are not greatly higher than the spring floodmarks of the present; showing conclusively that no extraordinary decrease has taken place in the volume of the upper tributaries of post-glacial Sierra streams since they came into existence. But in the mean time, eliminating all this complicated question of climatic change, the plain fact remains that *the present rain- and snowfall is abundantly sufficient for the luxuriant growth of Sequoia forests.* Indeed, all my observations tend to show that in a prolonged drought the Sugar Pines and firs would perish before the Sequoia, not alone because of the greater longevity of individual trees, but because the species can endure more drought, and make the most of whatever moisture falls.

Again, if the restriction and irregular distribution of the species be interpreted as a result of the desiccation of the range, then instead of increasing as it does in individuals toward the south where the rainfall is less, it should diminish.

If, then, the peculiar distribution of Sequoia has
not been governed by superior conditions of soil
as to fertility or moisture, by what has it been
governed?

In the course of my studies I observed that the
northern groves, the only ones I was at first acquainted with, were located on just those portions
of the general forest soil-belt that were first laid
bare toward the close of the glacial period when
the ice-sheet began to break up into individual
glaciers. And while searching the wide basin of
the San Joaquin, and trying to account for the
absence of Sequoia where every condition seemed
favorable for its growth, it occurred to me that this
remarkable gap in the Sequoia belt is located exactly in the basin of the vast ancient *mer de glace*
of the San Joaquin and King's River basins, which
poured its frozen floods to the plain, fed by the
snows that fell on more than fifty miles of the
summit. I then perceived that the next great gap
in the belt to the northward, forty miles wide, extending between the Calaveras and Tuolumne
groves, occurs in the basin of the great ancient *mer
de glace* of the Tuolumne and Stanislaus basins,
and that the smaller gap between the Merced and
Mariposa groves occurs in the basin of the smaller
glacier of the Merced. *The wider the ancient glacier,
the wider the corresponding gap in the Sequoia belt.*

Finally, pursuing my investigations across the
basins of the Kaweah and Tule, I discovered that
the Sequoia belt attained its greatest development
just where, owing to the topographical peculiari-
ties of the region, the ground had been most per-

fectly protected from the main ice-rivers that continued to pour past from the summit fountains long after the smaller local glaciers had been melted.

Taking now a general view of the belt, beginning at the south, we see that the majestic ancient glaciers were shed off right and left down the valleys of Kern and King's rivers by the lofty protective spurs outspread embracingly above the warm Sequoia-filled basins of the Kaweah and Tule. Then, next northward, occurs the wide Sequoia-less channel, or basin, of the ancient San Joaquin and King's River *mer de glace;* then the warm, protected spots of Fresno and Mariposa groves; then the Sequoia-less channel of the ancient Merced glacier; next the warm, sheltered ground of the Merced and Tuolumne groves; then the Sequoia-less channel of the grand ancient *mer de glace* of the Tuolumne and Stanislaus; then the warm old ground of the Calaveras and Stanislaus groves. It appears, therefore, that just where, at a certain period in the history of the Sierra, the glaciers were not, there the Sequoia is, and just where the glaciers were, there the Sequoia is not.

What the other conditions may have been that enabled Sequoia to establish itself upon these oldest and warmest portions of the main glacial soil-belt, I cannot say. I might venture to state, however, in this connection, that since the Sequoia forests present a more and more ancient aspect as they extend southward, I am inclined to think that the species was distributed from the south, while the Sugar Pine, its great rival in the northern groves, seems to have come around the head of the Sacramento

VIEW IN TUOLUMNE CAÑON, YOSEMITE NATIONAL PARK.

valley and down the Sierra from the north; consequently, when the Sierra soil-beds were first thrown open to preëmption on the melting of the ice-sheet, the Sequoia may have established itself along the available portions of the south half of the range prior to the arrival of the Sugar Pine, while the Sugar Pine took possession of the north half prior to the arrival of Sequoia.

But however much uncertainty may attach to this branch of the question, there are no obscuring shadows upon the grand general relationship we have pointed out between the present distribution of Sequoia and the ancient glaciers of the Sierra. And when we bear in mind that all the present forests of the Sierra are young, growing on moraine soil recently deposited, and that the flank of the range itself, with all its landscapes, is new-born, recently sculptured, and brought to the light of day from beneath the ice mantle of the glacial winter, then a thousand lawless mysteries disappear, and broad harmonies take their places.

But although all the observed phenomena bearing on the post-glacial history of this colossal tree point to the conclusion that it never was more widely distributed on the Sierra since the close of the glacial epoch; that its present forests are scarcely past prime, if, indeed, they have reached prime; that the post-glacial day of the species is probably not half done; yet, when from a wider outlook the vast antiquity of the genus is considered, and its ancient richness in species and individuals; comparing our Sierra Giant and *Sequoia sempervirens* of the Coast Range, the only other living species of Sequoia, with the twelve fossil species already discovered and described by Heer and Lesquereux, some of which seem to have flourished over vast areas in the Arctic regions and in Europe and our own territories, during tertiary and cretaceous times,—then indeed it becomes plain that our two surviving species, restricted to narrow belts within the limits of California, are mere remnants of the genus, both as to species and individuals, and that they probably are verging to extinction. But the verge of a period beginning in cretaceous times may have a breadth of tens of thousands of years, not to mention the possible existence of conditions calculated to multiply and reëxtend both species and individuals. This, however, is a branch of the question into which I do not now purpose to enter.

In studying the fate of our forest king, we have thus far considered the action of purely natural causes only; but, unfortunately, *man* is in the woods, and waste and pure destruction are making rapid headway. If the importance of forests were at all understood, even from an economic standpoint, their preservation would call forth the most watchful attention of government. Only of late years by means of forest reservations has the simplest groundwork for available legislation been laid, while in many of the finest groves every species of destruction is still moving on with accelerated speed.

In the course of my explorations I found no fewer than five mills located on or near the lower edge of the Sequoia belt, all of which were cutting considerable quantities of Big Tree lumber.

Most of the Fresno group are doomed to feed the mills recently erected near them, and a company of lumbermen are now cutting the magnificent forest on King's River. In these milling operations waste far exceeds use, for after the choice young manageable trees on any given spot have been felled, the woods are fired to clear the ground of limbs and refuse with reference to further operations, and, of course, most of the seedlings and saplings are destroyed.

These mill ravages, however, are small as compared with the comprehensive destruction caused by "sheepmen." Incredible numbers of sheep are driven to the mountain pastures every summer, and their course is ever marked by desolation. Every wild garden is trodden down, the shrubs are stripped of leaves as if devoured by locusts, and the woods are burned. Running fires are set everywhere, with a view to clearing the ground of prostrate trunks, to facilitate the movements of the flocks and improve the pastures. The entire forest belt is thus swept and devastated from one extremity of the range to the other, and, with the exception of the resinous *Pinus contorta*, Sequoia suffers most of all. Indians burn off the underbrush in certain localities to facilitate deer-hunting, mountaineers and lumbermen carelessly allow their camp-fires to run; but the fires of the sheepmen, or *muttoneers*, form more than ninety per cent. of all destructive fires that range the Sierra forests.

It appears, therefore, that notwithstanding our forest king might live on gloriously in Nature's keeping, it is rapidly vanishing before the fire and steel of man; and unless protective measures be speedily invented and applied, in a few decads, at the farthest, all that will be left of *Sequoia gigantea* will be a few hacked and scarred monuments.

TWO-LEAVED, OR TAMARACK, PINE
(*Pinus contorta*, var. *Marrayana*)

THIS species forms the bulk of the alpine forests, extending along the range, above the fir zone, up to a height of from 8000 to 9500 feet above the sea, growing in beautiful order upon moraines that are scarcely changed as yet by post-glacial weathering. Compared with the giants of the lower zones, this is a small tree, seldom attaining a height of a hundred feet. The largest specimen I ever measured was ninety feet in height, and a little over six in diameter four feet from the ground. The average height of mature trees throughout the entire belt is probably not far from fifty or sixty feet, with a diameter of two feet. It is a well-proportioned, rather handsome little pine, with grayish-brown bark, and crooked, much-divided branches, which cover the greater portion of the trunk, not so densely, however, as to prevent its being seen. The lower limbs curve downward, gradually take a horizontal position about half-way up the trunk, then aspire more and more toward the summit, thus forming a sharp, conical top. The foliage is short and rigid, two leaves in a fascicle, arranged in comparatively long, cylindrical tassels at the ends of the tough, upcurving branchlets. The cones are about two inches long, growing in stiff clusters among the needles, without making any striking effect, except while very young, when they are of a vivid crimson color, and the whole tree appears to be dotted with brilliant flowers. The sterile cones are still more showy, on account of their great abundance, often giving a reddish-yellow tinge to the whole mass of the foliage, and filling the air with pollen.

No other pine on the range is so regularly planted as this one. Moraine forests sweep along the sides of the high, rocky valleys for miles without interruption; still, strictly speaking, they are not dense, for flecks of sunshine and flowers find their way into the darkest places, where the trees grow tallest and thickest. Tall, nutritious grasses are specially abundant beneath them, growing over all the ground, in sunshine and shade, over extensive areas like a farmer's crop, and serving as pasture for the multitude of sheep that are driven from the arid plains every summer as soon as the snow is melted.

The Two-leaved Pine, more than any other, is subject to destruction by fire. The thin bark is streaked and sprinkled with resin, as though it had been showered down upon it like rain, so that even the green trees catch fire readily, and during strong winds whole forests are destroyed, the flames leaping from tree to tree, forming one continuous belt of roaring fire that goes surging and racing onward above the bending woods, like the grass-fires of a prairie. During the calm, dry season of Indian summer, the fire creeps quietly along the ground, feeding on the dry needles and burs; then, arriving at the foot of a tree, the resiny bark is ignited, and the heated air ascends in a powerful current, increasing in velocity, and dragging the flames swiftly upward; then the leaves catch fire, and an immense column of flame, beautifully spired on the edges, and tinted a rose-purple hue, rushes aloft thirty or forty feet above the top of the tree, forming a grand spectacle, especially on a dark night. It lasts, however, only a few seconds, vanishing with magical rapidity, to be succeeded by others along the fire-line at irregular intervals for weeks at a time — tree after tree flashing and darkening, leaving the trunks and branches hardly scarred. The heat, however, is sufficient to kill the trees, and in a few years the bark shrivels and falls off. Belts miles in extent are thus killed and left standing with the branches on, peeled and rigid, appearing gray in the distance, like misty clouds. Later the branches drop off, leaving a forest of bleached spars. At length the roots decay, and the forlorn trunks are blown down during some storm, and piled one upon another encumbering the ground until they are consumed by the next fire, and leave it ready for a fresh crop.

The endurance of the species is shown by its wandering occasionally out over the lava plains with the Yellow Pine, and climbing moraineless mountain-sides with the Dwarf Pine, clinging to any chance support in rifts and crevices of storm-beaten rocks—always, however, showing the effects of such hardships in every feature.

Down in sheltered lake hollows, on beds of rich alluvium, it varies so far from the common form as frequently to be taken for a distinct species. Here it grows in dense sods, like grasses, from forty to eighty feet high, bending all together to the breeze and whirling in eddying gusts more lithely than any other tree in the woods. I have frequently found specimens fifty feet high less than five inches in diameter. Being thus slender, and at the same time well clad with leafy boughs, it is oftentimes bent to the ground when laden with soft snow, forming beautiful arches in endless variety, some of which last until the melting of the snow in spring.

MOUNTAIN PINE

(*Pinus monticola*)

THE Mountain Pine is king of the alpine woods, brave, hardy, and long-lived, towering grandly above its companions, and becoming stronger and more imposing just where other species begin to crouch and disappear. At its best it is usually about ninety feet high and five or six in diameter, though a specimen is often met considerably larger than this. The trunk is as massive and as suggestive of enduring strength as that of an oak. About two thirds of the trunk is commonly free of limbs, but close, fringy tufts of sprays occur all the way down, like those which adorn the colossal shafts of Sequoia. The bark is deep reddish-brown upon trees that occupy exposed situations near its upper limit, and furrowed rather deeply, the main furrows running nearly parallel with each other, and connected by conspicuous cross furrows, which, with one exception, are, as far as I have noticed, peculiar to this species.

The cones are from four to eight inches long, slender, cylindrical, and somewhat curved, resembling those of the common White Pine of the Atlantic coast. They grow in clusters of about from three to six or seven, becoming pendulous as they increase in weight, chiefly by the bending of the branches.

This species is nearly related to the Sugar Pine, and, though not half so tall, it constantly suggests its noble relative in the way that it extends its long arms and in general habit. The Mountain Pine is first met on the upper margin of the fir zone, growing singly in a subdued, inconspicuous form, in what appear as chance situations, without making much impression on the general forest. Continuing up through the Two-leaved Pines in the same scattered growth, it begins to show its character, and at an elevation of about 10,000 feet attains its noblest development near the middle of the range, tossing its tough arms in the frosty air, welcoming storms and feeding on them, and reaching the grand old age of 1000 years.

JUNIPER, OR RED CEDAR

(*Juniperus occidentalis*)

THE Juniper is preëminently a rock tree, occupying the baldest domes and pavements, where there is scarcely a handful of soil, at a height of from 7000 to 9500 feet. In such situations the trunk is frequently over eight feet in diameter, and not much more in height. The top is almost always dead in old trees, and great stubborn limbs push

out horizontally that are mostly broken and bare
at the ends, but densely covered and embedded
here and there with bossy mounds of gray foliage.
Some are mere weathered stumps, as broad as long,
decorated with a few leafy sprays, reminding one
of the crumbling towers of some ancient castle

JUNIPER, OR RED CEDAR.

scantily draped with ivy. Only upon the head
waters of the Carson have I found this species established on good moraine soil. Here it flourishes
with the Silver and Two-leaved Pines, in great
beauty and luxuriance, attaining a height of from
forty to sixty feet, and manifesting but little of
that rocky angularity so characteristic a feature

throughout the greater portion of its range. Two of the largest, growing at the head of Hope Valley, measured twenty-nine feet three inches and twenty-five feet six inches in circumference, respectively, four feet from the ground.   The bark is of a bright cinnamon color, and, in thrifty trees, beautifully braided and reticulated, flaking off in thin, lustrous ribbons that are sometimes used by Indians for tent-matting. Its fine color and odd picturesqueness always catch an artist's eye, but to me the Juniper seems a singularly dull and taciturn tree, never speaking to one's heart.   I have spent many a day and night in its company, in all kinds of weather, and have ever found it silent, cold, and rigid, like a column of ice.   Its broad stumpiness, of course, precludes all possibility of waving, or even shaking; but it is not this rocky steadfastness that constitutes its silence. In calm, sun-days the Sugar Pine preaches the grandeur of the mountains like an apostle without moving a leaf.

On level rocks it dies standing, and wastes insensibly out of existence like granite, the wind exerting about as little control over it alive or dead as it does over a glacier boulder. Some are undoubtedly over 2000 years old. All the trees of the alpine woods suffer, more or less, from avalanches, the Two-leaved Pine most of all. Gaps two or three hundred yards wide, extending from the upper limit of the tree-line to the bottoms of valleys and lake basins, are of common occurrence in all the upper forests, resembling the clearings of settlers in the old backwoods. Scarcely a

STORM-BEATEN JUNIPERS.

tree is spared, even the soil is scraped away, while the thousands of uprooted pines and spruces are piled upon one another heads downward, and tucked snugly in along the sides of the clearing in two windrows, like lateral moraines. The pines lie with branches wilted and drooping like weeds. Not so the burly junipers. After braving in silence the storms of perhaps a dozen or twenty centuries, they seem in this, their last calamity, to become somewhat communicative, making sign of a very unwilling acceptance of their fate, holding themselves well up from the ground on knees and elbows, seemingly ill at ease, and anxious, like stubborn wrestlers, to rise again.

HEMLOCK SPRUCE

(*Tsuga Pattoniana*)

THE Hemlock Spruce is the most singularly beautiful of all the California coniferæ. So slender is its axis at the top, that it bends over and droops like the stalk of a nodding lily. The branches droop also, and divide into innumerable slender, waving sprays, which are arranged in a varied, eloquent harmony that is wholly indescribable. Its cones are purple, and hang free, in the form of little tassels two inches long from all the sprays from top to bottom. Though exquisitely delicate and feminine in expression, it grows best where the snow lies deepest, far up in the region of storms, at an elevation of from 9000 to 9500 feet, on frosty northern slopes; but it is capable of

STORM-BEATEN HEMLOCK SPRUCE,
FORTY FEET HIGH.

growing considerably higher, say 10,500 feet. The tallest specimens, growing in sheltered hollows somewhat beneath the heaviest wind-currents, are from eighty to a hundred feet high, and from two to four feet in diameter. The very largest specimen I ever found was nineteen feet seven inches in circumference four feet from the ground, growing on the edge of Lake Hollow, at an elevation of 9250 feet above the level of the sea. At the age of twenty or thirty years it becomes fruitful, and hangs out its beautiful purple cones at the ends of the slender sprays, where they swing free in the breeze, and contrast delightfully with the cool green foliage. They are translucent when young, and their beauty is delicious. After they are fully ripe, they spread their shell-like scales and allow the brown-winged seeds to fly in the mellow air, while the empty cones remain to beautify the tree until the coming of a fresh crop.

The staminate cones of all the coniferæ are beautiful, growing in bright clusters, yellow, and rose, and crimson. Those of the Hemlock Spruce are the most beautiful of all, forming little conelets of blue flowers, each on a slender stem.

Under all conditions, sheltered or stormbeaten, well-fed or ill-fed, this tree is singularly graceful in habit. Even at its highest limit upon exposed ridge-tops, though compelled to crouch in dense thickets, huddled close together, as if for mutual protection, it still manages to throw out its sprays in irrepressible loveliness; while on well-ground moraine soil it develops a perfectly tropical luxuriance of foliage and fruit, and is the very loveliest tree in the forest; poised in thin white sunshine, clad with branches from head to foot, yet not in the faintest degree heavy or bunchy, it towers in unassuming majesty, drooping as if unaffected with the aspiring tendencies of its race, loving the ground while transparently conscious of heaven and joyously receptive of its blessings, reaching out its branches like sensitive tentacles, feeling the light and reveling in it. No other of our alpine conifers so finely veils its strength. Its delicate branches yield to the mountains' gentlest breath; yet is it strong to meet the wildest onsets of the gale,—strong not in resistance, but compliance, bowing, snow-laden, to the ground, gracefully accepting burial month after month in the darkness beneath the heavy mantle of winter.

14

When the first soft snow begins to fall, the flakes lodge in the leaves, weighing down the branches against the trunk. Then the axis bends yet lower and lower, until the slender top touches the ground, thus forming a fine ornamental arch. The snow still falls lavishly, and the whole tree is at length buried, to sleep and rest in its beautiful grave as though dead. Entire groves of young trees, from ten to forty feet high, are thus buried every winter like slender grasses. But, like the violets and daisies which the heaviest snows crush not, they are safe. It is as though this were only Nature's method of putting her darlings to sleep instead of leaving them exposed to the biting storms of winter.

Thus warmly wrapped they await the summer resurrection. The snow becomes soft in the sunshine, and freezes at night, making the mass hard and compact, like ice, so that during the months of April and May you can ride a horse over the prostrate groves without catching sight of a single leaf. At length the down-pouring sunshine sets them free. First the elastic tops of the arches begin to appear, then one branch after another, each springing loose with a gentle rustling sound, and at length the whole tree, with the assistance of the winds, gradually unbends and rises and settles back into its place in the warm air, as dry and feathery and fresh as young ferns just out of the coil.

Some of the finest groves I have yet found are on the southern slopes of Lassen's Butte. There are also many charming companies on the head waters of the Tuolumne, Merced, and San Joaquin, and, in general, the species is so far from being rare that you can scarcely fail to find groves of considerable extent in crossing the range, choose what pass you may. The Mountain Pine grows beside it, and more frequently the two-leaved species; but there are many beautiful groups, numbering 1000 individuals, or more, without a single intruder.

I wish I had space to write more of the surpassing beauty of this favorite spruce. Every tree-lover is sure to regard it with special admiration; apathetic mountaineers, even, seeking only game or gold, stop to gaze on first meeting it, and mutter to themselves: "That's a mighty pretty tree," some of them adding, "d——d pretty!" In autumn, when its cones are ripe, the little striped tamias, and the Douglas squirrel, and the Clark crow make a happy stir in its groves. The deer love to lie down beneath its spreading branches; bright streams from the snow that is always near ripple through its groves, and bryanthus spreads precious carpets in its shade. But the best words only hint its charms. Come to the mountains and see.

DWARF PINE

(*Pinus albicaulis*)

THIS species forms the extreme edge of the timber line throughout nearly the whole extent of the range on both flanks. It is first met growing in company with *Pinus contorta*, var. *Murrayana*, on the upper margin of the belt, as an erect tree from fifteen to thirty feet high and from one to two feet in thickness; thence it goes straggling up the flanks of the summit peaks, upon moraines or crumbling ledges, wherever it can obtain a foothold, to an elevation of from 10,000 to 12,000 feet, where it dwarfs to a mass of crumpled, prostrate branches, covered with slender, upright shoots, each tipped with a short, close-packed tassel of leaves.   The bark is smooth

GROUP OF ERECT DWARF PINES.

and purplish, in some places almost white. The fertile cones grow in rigid clusters upon the upper branches, dark chocolate in color while young, and bear beautiful pearly seeds about the size of peas, most of which are eaten by two species of tamias and the notable Clark crow.   The staminate cones occur in clusters, about an inch wide, down among the leaves, and, as they are colored bright rose-purple, they give rise to a lively, flowery appearance little looked for in such a tree.

Pines are commonly regarded as sky-loving trees that must necessarily aspire or die. This species forms a marked exception, creeping lowly, in compliance with the most rigorous demands of climate, yet enduring bravely to a more advanced age than many of its lofty relatives in the sun-lands below. Seen from a distance, it would never be taken for a tree of any kind. Yonder, for example, is Cathedral Peak, some three miles away, with a scattered growth of this pine creeping like mosses over the roof and around the beveled edges of the north gable, nowhere giving any hint of an ascending axis. When approached quite near it still appears matted and heathy, and is so low that one experiences no great difficulty in walking over the top of it. Yet it is seldom absolutely prostrate, at its lowest usually attaining a height of three or four feet, with a main trunk, and branches outspread and intertangled above it, as if in ascending they had been checked by a ceiling, against which they had grown and been compelled to spread horizontally. The winter snow is indeed such a ceiling, lasting half the year; while the pressed, shorn surface is made yet smoother by violent winds, armed with cutting sand-grains, that beat down any shoot that offers to rise much above the general level, and carve the dead trunks and branches in beautiful patterns.

During stormy nights I have often camped snugly beneath the interlacing arches of this little

pine. The needles, which have accumulated for
centuries, make fine beds, a fact well known to other

A DWARF PINE.

mountaineers, such as deer and wild sheep, who
paw out oval hollows and lie beneath the larger
trees in safe and comfortable concealment.

The longevity of this lowly dwarf is far greater than would be guessed. Here, for example, is a specimen, growing at an elevation of 10,700 feet, which seems as though it might be plucked up by the roots, for it is only three and a half inches in diameter, and its topmost tassel is hardly three feet above the ground. Cutting it half through and counting the annual rings with the aid of a lens, we find its age to be no less than 255 years. Here is another telling specimen about the same height, 426 years old, whose trunk is only six inches in diameter; and one of its supple branchlets, hardly an eighth of an inch in diameter inside the bark, is seventy-five years old, and so filled with oily balsam, and so well seasoned by storms, that we may tie it in knots like a whip-cord.

WHITE PINE
(*Pinus flexilis*)

THIS species is widely distributed throughout the Rocky Mountains, and over all the higher of the many ranges of the Great Basin, between the Wahsatch Mountains and the Sierra, where it is known as White Pine. In the Sierra it is sparsely scattered along the eastern flank, from Bloody Cañon southward nearly to the extremity of the range, opposite the village of Lone Pine, nowhere forming any appreciable portion of the general forest. From its peculiar position, in loose, straggling parties, it seems to have been derived from the Basin ranges to the eastward, where it is abundant.

It is a larger tree than the Dwarf Pine. At an elevation of about 9000 feet above the sea, it often attains a height of forty or fifty feet, and a diameter of from three to five feet. The cones open freely when ripe, and are twice as large as those of the *albicaulis*, and the foliage and branches are more open, having a tendency to sweep out in free, wild curves, like those of the Mountain Pine, to which it is closely allied. It is seldom found lower than 9000 feet above sea-level, but from this elevation it pushes upward over the roughest ledges to the extreme limit of tree-growth, where, in its dwarfed, storm-crushed condition, it is more like the white-barked species.

Throughout Utah and Nevada it is one of the principal timber-trees, great quantities being cut every year for the mines. The famous White Pine Mining District, White Pine City, and the White Pine Mountains have derived their names from it.

NEEDLE PINE

(*Pinus aristata*)

THIS species is restricted in the Sierra to the southern portion of the range, about the head waters of Kings and Kern rivers, where it forms extensive forests, and in some places accompanies the Dwarf Pine to the extreme limit of tree-growth.

It is first met at an elevation of between 9000 and 10,000 feet, and runs up to 11,000 without seeming to suffer greatly from the climate or the leanness of the soil. It is a much finer tree than the

Dwarf Pine. Instead of growing in clumps and low, heathy mats, it manages in some way to maintain an erect position, and usually stands single. Wherever the young trees are at all sheltered, they grow up straight and arrowy, with delicately tapered bole, and ascending branches terminated

OAK GROWING AMONG YELLOW PINES.

with glossy, bottle-brush tassels. At middle age, certain limbs are specialized and pushed far out for the bearing of cones, after the manner of the Sugar Pine; and in old age these branches droop and cast about in every direction, giving rise to very picturesque effects. The trunk becomes deep brown and rough, like that of the Mountain Pine, while the young cones are of a strange, dull, blackish-blue color, clustered on the upper branches.   When ripe they are from three to four inches long, yellowish brown, resembling in every way those of the Mountain Pine.   Excepting the Sugar Pine, no tree on the mountains is so capable of individual expression, while in grace of form and movement it constantly reminds one of the Hemlock Spruce.

The largest specimen I measured was a little over five feet in diameter and ninety feet in height, but this is more than twice the ordinary size.

This species is common throughout the Rocky Mountains and most of the short ranges of the Great Basin, where it is called the Fox-tail Pine, from its long dense leaf-tassels.   On the Hot Creek, White Pine, and Golden Gate ranges it is quite abundant.   About a foot or eighteen inches of the ends of the branches is densely packed with stiff outstanding needles which radiate like an electric fox or squirrel's tail.   The needles have a glossy polish, and the sunshine sifting through them makes them burn with silvery luster, while their number and elastic temper tell delightfully in the winds. This tree is here still more original and picturesque than in the Sierra, far surpassing not only its companion conifers in this respect, but also the most noted of the lowland oaks.   Some stand firmly erect, feathered with radiant tassels down to the ground, forming slender tapering towers of shining verdure; others, with two or three specialized branches pushed out at right angles to the trunk and densely clad with tasseled sprays, take the form of beautiful ornamental crosses.   Again in the same woods you

find trees that are made up of several boles united
near the ground, spreading at the sides in a plane
parallel to the axis of the mountain, with the elegant tassels hung in charming order between them,
making a harp held against the main wind lines
where they are most effective in playing the grand
storm harmonies. And besides these there are
many variable arching forms, alone or in groups,
with innumerable tassels drooping beneath the
arches or radiant above them, and many lowly
giants of no particular form that have braved the
storms of a thousand years. But whether old or
young, sheltered or exposed to the wildest gales,
this tree is ever found irrepressibly and extravagantly picturesque, and offers a richer and more
varied series of forms to the artist than any other
conifer I know of.

NUT PINE

(*Pinus monophylla*)

THE Nut Pine covers or rather dots the eastern
flank of the Sierra, to which it is mostly restricted,
in grayish, bush-like patches, from the margin of the
sage-plains to an elevation of from 7000 to 8000 feet.

A more contentedly fruitful and unaspiring conifer could not be conceived. All the species we
have been sketching make departures more or less
distant from the typical spire form, but none goes
so far as this. Without any apparent exigency of
climate or soil, it remains near the ground, throwing
out crooked, divergent branches like an orchard

apple-tree, and seldom pushes a single shoot higher than fifteen or twenty feet above the ground.

The average thickness of the trunk is, perhaps, about ten or twelve inches. The leaves are mostly undivided, like round awls, instead of being separated, like those of other pines, into twos and threes and fives. The cones are green while growing, and are usually found over all the tree, forming quite a marked feature as seen against the bluish-gray foliage. They are quite small, only about two inches in length, and give no promise of edible nuts; but when we come to open them, we find that about half the entire bulk of the cone is made up of sweet, nutritious seeds, the kernels of which are nearly as large as those of hazel-nuts.

This is undoubtedly the most important food-tree on the Sierra, and furnishes the Mono, Carson, and Walker River Indians with more and better nuts than all the other species taken together. It is the Indians' own tree, and many a white man have they killed for cutting it down.

In its development Nature seems to have aimed at the formation of as great a fruit-bearing surface as possible. Being so low and accessible, the cones are readily beaten off with poles, and the nuts procured by roasting them until the scales open. In bountiful seasons a single Indian will gather thirty or forty bushels of them — a fine squirrelish employment.

Of all the conifers along the eastern base of the Sierra, and on all the many mountain groups and short ranges of the Great Basin, this foodful little pine is the commonest tree, and the most important. Nearly every mountain is planted with it to a height of from 8000 to 9000 feet above the sea. Some are covered from base to summit by this one species, with only a sparse growth of juniper on the lower slopes to break the continuity of its curious woods, which, though dark-looking at a distance, are almost shadeless, and have none of the damp, leafy glens and hollows so characteristic of other pine woods. Tens of thousands of acres occur in continuous belts. Indeed, viewed comprehensively the entire Basin seems to be pretty evenly divided into level plains dotted with sage-bushes and mountain-chains covered with Nut Pines. No slope is too rough, none too dry, for these bountiful orchards of the red man.

The value of this species to Nevada is not easily overestimated. It furnishes charcoal and timber for the mines, and, with the juniper, supplies the ranches with fuel and rough fencing. In fruitful seasons the nut crop is perhaps greater than the California wheat crop, which exerts so much influence throughout the food markets of the world. When the crop is ripe, the Indians make ready the long beating-poles; bags, baskets, mats, and sacks are collected; the women out at service among the settlers, washing or drudging, assemble at the family huts; the men leave their ranch work; old and young, all are mounted on ponies and start in great glee to the nut-lands, forming curiously picturesque cavalcades; flaming scarfs and calico skirts stream loosely over the knotty ponies, two squaws usually astride of each, with baby midgets bandaged in baskets slung on their backs or balanced on the saddle-bow; while nut-baskets and water-jars project from each side, and the long beating-poles make angles in every direction. Arriving at some well-known central point where grass and water are found, the squaws with baskets, the men with poles ascend the ridges to the laden trees, followed by the children. Then the beating begins right merrily, the burs fly in every direction, rolling down the slopes, lodging here and there against rocks and sage-bushes, chased and gathered by the women and children with fine natural gladness. Smoke-columns speedily mark the joyful scene of their labors as the roasting-fires are kindled, and, at night, assembled in gay circles garrulous as jays, they begin the first nut feast of the season.

The nuts are about half an inch long and a quarter of an inch in diameter, pointed at the top, round at the base, light brown in general color, and, like many other pine seeds, handsomely dotted with purple, like birds' eggs. The shells are thin and may be crushed between the thumb and finger. The kernels are white, becoming brown by roasting, and are sweet to every palate, being eaten by birds, squirrels, dogs, horses, and men. Perhaps less than one bushel in a thousand of the whole crop is ever gathered. Still, besides supplying their own wants, in times of plenty the Indians bring large quantities to market; then they are eaten around nearly every fireside in the State, and are even fed to horses occasionally instead of barley.

Of other trees growing on the Sierra, but forming a very small part of the general forest, we may briefly notice the following:

Chamæcyparis Lawsoniana is a magnificent tree in the coast ranges, but small in the Sierra. It is found only well to the northward along the banks of cool streams on the upper Sacramento toward Mount Shasta. Only a few trees of this species, as far as I have seen, have as yet gained a place in the Sierra woods. It has evidently been derived from the coast range by way of the tangle of connecting mountains at the head of the Sacramento Valley.

In shady dells and on cool stream banks of the northern Sierra we also find the Yew (*Taxus brevifolia*).

The interesting Nutmeg Tree (*Torreya Californica*) is sparsely distributed along the western flank of the range at an elevation of about 4000 feet, mostly in gulches and cañons. It is a small, prickly leaved, glossy evergreen, like a conifer, from twenty to fifty feet high, and one to two feet in diameter. The fruit resembles a green-gage plum, and contains one seed, about the size of an acorn, and like a nutmeg, hence the common name. The wood is fine-grained and of a beautiful, creamy yellow color like box, sweet-scented when dry, though the green leaves emit a disagreeable odor.

Betula occidentalis, the only birch, is a small, slender tree restricted to the eastern flank of the range along stream-sides below the pine-belt, especially in Owen's Valley.

Alder, Maple, and Nuttall's Flowering Dogwood make beautiful bowers over swift, cool streams at an elevation of from 3000 to 5000 feet, mixed more or less with willows and cottonwood; and above these in lake basins the aspen forms fine ornamental groves, and lets its light shine gloriously in the autumn months.

The Chestnut Oak (*Quercus densiflora*) seems to have come from the coast range around the head of the Sacramento Valley, like the *Chamæcyparis*, but as it extends southward along the lower edge of the main pine-belt it grows smaller until it finally dwarfs to a mere chaparral bush. In the coast mountains it is a fine, tall, rather slender tree, about from sixty to seventy-five feet high, growing with the grand *Sequoia sempervirens*, or Redwood. But unfortunately it is too good to live, and is now being rapidly destroyed for tanbark.

Besides the common Douglas Oak and the grand *Quercus Wislizeni* of the foot-hills, and several small ones that make dense growths of chaparral, there are two mountain-oaks that grow with the pines up to an elevation of about 5000 feet above the sea, and greatly enhance the beauty of the yosemite parks. These are the Mountain Live Oak and the Kellogg Oak, named in honor of the admirable botanical pioneer of California. Kellogg's Oak (*Quercus Kelloggii*) is a firm, bright, beautiful tree, reaching a height of sixty feet, four to seven feet in diameter, with wide-spreading branches, and growing at an elevation of from 3000 to 5000 feet in sunny valleys and flats among the evergreens, and higher in a dwarfed state. In the cliff-bound parks about 4000 feet above the sea it is so abundant and effective it might fairly be called the Yosemite Oak. The leaves make beautiful

PATE VALLEY, SHOWING THE OAKS—TUOLUMNE CAÑON, YOSEMITE NATIONAL PARK.

masses of purple in the spring, and yellow in ripe autumn; while its acorns are eagerly gathered by Indians, squirrels, and woodpeckers. The Mountain Live Oak (*Q. Chrysolepis*) is a tough, rugged mountaineer of a tree, growing bravely and attaining noble dimensions on the roughest earthquake taluses in deep cañons and yosemite valleys. The trunk is usually short, dividing near the ground into great, wide-spreading limbs, and these again into a multitude of slender sprays, many of them cord-like and drooping to the ground, like those of the Great White Oak of the lowlands (*Q. lobata*). The top of the tree where there is plenty of space is broad and bossy, with a dense covering of shining leaves, making delightful canopies, the complicated system of gray, interlacing, arching branches as seen from beneath being exceedingly rich and picturesque. No other tree that I know dwarfs so regularly and completely as this under changes of climate due to changes in elevation. At the foot of a cañon 4000 feet above the sea you may find magnificent specimens of this oak fifty feet high, with craggy, bulging trunks, five to seven feet in diameter, and at the head of the cañon, 2500 feet higher, a dense, soft, low, shrubby growth of the same species, while all the way up the cañon between these extremes of size and habit a perfect gradation may be traced. The largest I have seen was fifty feet high, eight feet in diameter, and about seventy-five feet in spread. The trunk was all knots and buttresses, gray like granite, and about as angular and irregular as the boulders on which it was growing—a type of steadfast, unwedgeable strength.

15

CHAPTER IX

THE Douglas Squirrel is by far the most interesting and influential of the California sciuridæ, surpassing every other species in force of character, numbers, and extent of range, and in the amount of influence he brings to bear upon the health and distribution of the vast forests he inhabits.

Go where you will throughout the noble woods of the Sierra Nevada, among the giant pines and spruces of the lower zones, up through the towering Silver Firs to the storm-bent thickets of the summit peaks, you everywhere find this little squirrel the master-existence. Though only a few inches long, so intense is his fiery vigor and restlessness, he stirs every grove with wild life, and makes himself more important than even the huge bears that shuffle through the tangled underbrush beneath him. Every wind is fretted by his voice, almost every bole and branch feels the sting of his sharp feet. How much the growth of the trees is stimulated by this means it is not easy to learn, but his action in manipulating their seeds is more appreciable. Nature has made him master forester and committed most of her coniferous crops to his paws. Probably over fifty per cent. of all the cones ripened on the Sierra are cut off and handled by the Douglas alone, and of those of the Big Trees perhaps ninety per cent. pass through his hands: the greater portion is of course stored away for food to last during the winter and spring, but some of them are tucked separately into loosely covered holes, where some of the seeds germinate and become trees. But the Sierra is only one of the many provinces over which he holds sway, for his dominion extends over all the Redwood Belt of the Coast Mountains, and far northward throughout the majestic forests of Oregon, Washington, and British Columbia. I make haste to mention these facts, to show upon how substantial a foundation the importance I ascribe to him rests.

The Douglas is closely allied to the Red Squirrel or Chickaree of the eastern woods. Ours may be a lineal descendant of this species, distributed westward to the Pacific by way of the Great Lakes and the Rocky Mountains, and thence southward along our forested ranges. This view is suggested by the fact that our species becomes redder and more Chickaree-like in general, the farther it is traced back along the course indicated above. But whatever their relationship, and the evolutionary forces that have acted upon them, the Douglas is now the larger and more beautiful animal.

From the nose to the root of the tail he measures about eight inches; and his tail, which he so effectively uses in interpreting his feelings, is about six inches in length. He wears dark bluish-gray over the back and half-way down the sides, bright buff on the belly, with a stripe of dark gray, nearly black, separating the upper and under colors; this dividing stripe, however, is not very sharply defined. He has long black whiskers, which gives him a rather fierce look when observed closely, strong claws, sharp as fish-hooks, and the brightest of bright eyes, full of telling speculation.

A King's River Indian told me that they call him "Pillillooeet," which, rapidly pronounced with the first syllable heavily accented, is not unlike the lusty exclamation he utters on his way up a tree when excited. Most mountaineers in California call him the Pine Squirrel; and when I asked an old trapper whether he knew our little forester, he replied with brightening countenance: "Oh, yes, of course I know him; everybody knows him. When I 'm huntin' in the woods, I often find out where the deer are by his barkin' at 'em. I call 'em Lightnin' Squirrels, because they 're so mighty quick and peert."

All the true squirrels are more or less birdlike in speech and movements; but the Douglas is preëminently so, possessing, as he does, every attribute peculiarly squirrelish enthusiastically concentrated. He is the squirrel of squirrels, flashing from branch to branch of his favorite evergreens crisp and glossy and undiseased as a sunbeam. Give him wings and he would outfly any bird in the woods. His big gray cousin is a looser animal, seemingly light enough to float on the wind; yet when leaping from limb to limb, or out of one tree-top to another, he sometimes halts to gather strength, as if making efforts concerning the upshot of which he does not always feel exactly confident. But the Douglas, with his denser body, leaps and glides in hidden strength, seemingly as independent of common muscles as a mountain stream. He threads the tasseled branches of the pines, stirring their needles like a rustling breeze; now shooting across openings in arrowy lines; now launching in curves, glinting deftly from side to side in sudden zigzags, and swirling in giddy loops and spirals around the knotty trunks; getting into what seem to be the most impossible situations without sense of danger; now on his haunches, now on his head; yet ever graceful, and punctuating his most irrepressible outbursts of energy with little dots and dashes of perfect repose. He is, without exception, the wildest animal I ever saw,— a fiery, sputtering little bolt of life, luxuriating in quick oxygen and the woods' best juices. One can hardly think of such a creature being dependent, like the rest of us, on climate and food. But, after all, it requires no long acquaintance to learn he is human, for he works for a living. His busiest time is in the Indian summer. Then he gathers burs and hazelnuts like a plodding farmer, working continuously every day for hours; saying not a word; cutting off the ripe cones at the top of his speed, as if employed by the job, and examining every branch in regular order, as if careful that not one should escape him; then, descending, he stores them away beneath logs and stumps, in anticipation of the pinching hunger days of winter. He seems himself a kind of coniferous fruit,— both fruit and flower. The resiny essences of the pines pervade every pore of his body, and eating his flesh is like chewing gum.

One never tires of this bright chip of nature,—this brave little voice crying in the wilderness,— of observing his many works and ways, and listening to his curious language. His musical, piny gossip is as savory to the ear as balsam to the palate; and, though he has not exactly the gift of song, some of his notes are as sweet as those of a linnet —almost flute-like in softness, while others prick and tingle like thistles. He is the mocking-bird of squirrels, pouring forth mixed chatter and song like a perennial fountain; barking like a dog, screaming like a hawk, chirping like a blackbird or a sparrow; while in bluff, audacious noisiness he is a very jay.

In descending the trunk of a tree with the intention of alighting on the ground, he preserves a cautious silence, mindful, perhaps, of foxes and wildcats; but while rocking safely at home in the pine-tops there is no end to his capers and noise; and woe to the gray squirrel or chipmunk that ventures to set foot on his favorite tree! No matter how slyly they trace the furrows of the bark, they are speedily discovered, and kicked down-stairs with comic vehemence, while a torrent of angry notes comes rushing from his whiskered lips that sounds remarkably like swearing. He will even attempt at times to drive away dogs and men, especially if he has had no previous knowledge of them. Seeing a man for the first time, he approaches nearer and nearer, until within a few feet; then, with an angry outburst, he makes a

sudden rush, all teeth and eyes, as if about to eat
you up. But, finding that the big, forked animal
does n't scare, he prudently beats a retreat, and sets
himself up to reconnoiter on some overhanging
branch, scrutinizing every movement you make
with ludicrous solemnity. Gath-
ering courage, he ventures down
the trunk again, churring and
chirping, and jerking nervously
up and down in curious loops,
eyeing you all the time, as if
showing off and demanding
your admiration. Finally, growing calmer, he settles down in
a comfortable posture on some
horizontal branch commanding
a good view, and beats time with
his tail to a steady " Chee-up!
chee-up!" or, when somewhat
less excited, "Pee-ah!" with the
first syllable keenly accented,
and the second drawn out like
the scream of a hawk,—repeat-
ing this slowly and more em-
phatically at first, then gradu-

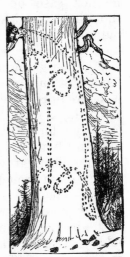

TRACK OF DOUGLAS
SQUIRREL ONCE DOWN
AND UP A PINE-TREE
WHEN SHOWING OFF TO
A SPECTATOR.

ally faster, until a rate of about 150 words a minute is reached; usually sitting all the time on his
haunches, with paws resting on his breast, which
pulses visibly with each word. It is remarkable,
too, that, though articulating distinctly, he keeps
his mouth shut most of the time, and speaks through
his nose. I have occasionally observed him even
eating Sequoia seeds and nibbling a troublesome

flea, without ceasing or in any way confusing his
"Pee-ah! pee-ah!" for a single moment.

While ascending trees all his claws come into
play, but in descending the weight of his body is
sustained chiefly by those of the hind feet; still in
neither case do his movements suggest effort,
though if you are near enough you may see the
bulging strength of his short, bear-like arms, and
note his sinewy fists clinched in the bark.

Whether going up or down, he carries his tail extended at full length in line with his body, unless
it be required for gestures. But while running
along horizontal limbs or fallen trunks, it is frequently folded forward over the back, with the airy
tip daintily upcurled. In cool weather it keeps him
warm. Then, after he has finished his meal, you
may see him crouched close on some level limb with
his tail-robe neatly spread and reaching forward to
his ears, the electric, outstanding hairs quivering
in the breeze like pine-needles. But in wet or very
cold weather he stays in his nest, and while curled
up there his comforter is long enough to come forward around his nose. It is seldom so cold, however, as to prevent his going out to his stores when
hungry.

Once as I lay storm-bound on the upper edge of
the timber line on Mount Shasta, the thermometer
nearly at zero and the sky thick with driving snow,
a Douglas came bravely out several times from one
of the lower hollows of a Dwarf Pine near my camp,
faced the wind without seeming to feel it much,
frisked lightly about over the mealy snow, and dug
his way down to some hidden seeds with wonder-

ful precision, as if to his eyes the thick snow-covering were glass.

No other of the Sierra animals of my acquaintance is better fed, not even the deer, amid abundance of sweet herbs and shrubs, or the mountain sheep, or omnivorous bears. His food consists of grass-seeds, berries, hazel-nuts, chinquapins, and the nuts and seeds of all the coniferous trees without exception,— Pine, Fir, Spruce, Libocedrus, Juniper, and Sequoia,— he is fond of them all, and they all agree with him, green or ripe. No cone is too large for him to manage, none so small as to be beneath his notice. The smaller ones, such as those of the Hemlock, and the Douglas Spruce, and the Two-leaved Pine, he cuts off and eats on a branch of the tree, without allowing them to fall; beginning at the bottom of the cone and cutting away the scales to expose the seeds; not gnawing by guess, like a bear, but turning them round and round in regular order, in compliance with their spiral arrangement.

When thus employed, his location in the tree is betrayed by a dribble of scales, shells, and seed-wings, and, every few minutes, by the fall of the stripped axis of the cone. Then of course he is ready for another, and if you are watching you may catch a glimpse of him as he glides silently out to the end of a branch and see him examining the cone-clusters until he finds one to his mind; then, leaning over, pull back the springy needles out of his way, grasp the cone with his paws to prevent its falling, snip it off in an incredibly short time, seize it with jaws grotesquely stretched, and return to his chosen seat near the trunk. But the immense size of the cones of the Sugar Pine— from fifteen to twenty inches in length—and those of the Jeffrey variety of the Yellow Pine compel him to adopt a quite different method. He cuts them off without attempting to hold them, then goes down and drags them from where they have chanced to fall up to the bare, swelling ground around the instep of the tree, where he demolishes them in the same methodical way, beginning at the bottom and following the scale-spirals to the top.

From a single Sugar Pine cone he gets from two to four hundred seeds about half the size of a hazel-

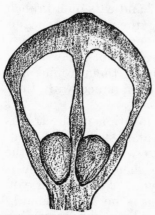

nut, so that in a few minutes he can procure enough to last a week. He seems, however, to prefer those of the two Silver First above all others; perhaps because they are most easily obtained, as the scales drop off when ripe without needing to be cut. Both species are filled with an exceedingly pungent, aromatic oil,

SEEDS, WINGS, AND SCALE OF
SUGAR PINE. (NAT. SIZE.)

which spices all his flesh, and is of itself sufficient

to account for his lightning energy.

You may easily know this little workman by his chips. On sunny hillsides around the principal trees they lie in big piles,—bushels and basketfuls of them, all fresh and clean, making the most beautiful kitchen-middens imaginable. The brown and yellow scales and nut-shells are as abundant and as delicately penciled and tinted as the shells along the sea-shore; while the beautiful red and purple seed-wings mingled with them would lead one to fancy that innumerable butterflies had there met their fate.

He feasts on all the species long before they are ripe, but is wise enough to wait until they are matured before he gathers them into his barns. This is in October and November, which with him are the two busiest months of the year. All kinds of burs, big and little, are now cut off and showered down alike, and the ground is speedily covered with them. A constant thudding and bumping is kept up; some of the larger cones chancing to fall on old logs make the forest reëcho with the sound. Other nut-eaters less industrious know well what is going on, and hasten to carry away the cones as they fall. But however busy the harvester may be, he is not slow to descry the pilferers below, and instantly leaves his work to drive them away. The little striped tamias is a thorn in his flesh, stealing persistently, punish him as he may. The large Gray Squirrel gives trouble also, although the Douglas has been accused of stealing from him. Generally, however, just the opposite is the case.

The excellence of the Sierra evergreens is well known to nurserymen throughout the world, consequently there is considerable demand for the seeds. The greater portion of the supply has hitherto been procured by chopping down the trees

in the more accessible sections of the forest alongside of bridle-paths that cross the range. Sequoia
seeds at first brought from twenty to thirty dollars
per pound, and therefore were eagerly sought after.
Some of the smaller fruitful trees were cut down in
the groves not protected by government, especially
those of Fresno and King's River. Most of the Sequoias, however, are of so gigantic a size that the
seedsmen have to look for the greater portion of
their supplies to the Douglas, who soon learns he is
no match for these freebooters. He is wise enough,
however, to cease working the instant he perceives
them, and never fails to embrace every opportunity
to recover his burs whenever they happen to be
stored in any place accessible to him, and the busy
seedsman often finds on returning to camp that
the little Douglas has exhaustively spoiled the
spoiler. I know one seed-gatherer who, whenever
he robs the squirrels, scatters wheat or barley beneath the trees as conscience-money.

The want of appreciable life remarked by so
many travelers in the Sierra forests is never felt
at this time of year. Banish all the humming insects and the birds and quadrupeds, leaving only
Sir Douglas, and the most solitary of our so-called
solitudes would still throb with ardent life. But
if you should go impatiently even into the most
populous of the groves on purpose to meet him,
and walk about looking up among the branches,
you would see very little of him. But lie down at
the foot of one of the trees and straightway he will
come. For, in the midst of the ordinary forest
sounds, the falling of burs, piping of quails, the

screaming of the Clark Crow, and the rustling of deer and bears among the chaparral, he is quick to detect your strange footsteps, and will hasten to make a good, close inspection of you as soon as you are still. First, you may hear him sounding a few notes of curious inquiry, but more likely the first intimation of his approach will be the prickly sounds of his feet as he descends the tree overhead, just before he makes his savage onrush to frighten you and proclaim your presence to every squirrel and bird in the neighborhood. If you remain perfectly motionless, he will come nearer and nearer, and probably set your flesh a-tingle by frisking across your body. Once, while I was seated at the foot of a Hemlock Spruce in one of the most inaccessible of the San Joaquin yosemites engaged in sketching, a reckless fellow came up behind me, passed under my bended arm, and jumped on my paper. And one warm afternoon, while an old friend of mine was reading out in the shade of his cabin, one of his Douglas neighbors jumped from the gable upon his head, and then with admirable assurance ran down over his shoulder and on to the book he held in his hand.

Our Douglas enjoys a large social circle; for, besides his numerous relatives, *Sciurus fossor, Tamias quadrivitatus, T. Townsendii, Spermophilus Beecheyi, S. Douglasii,* he maintains intimate relations with the nut-eating birds, particularly the Clark Crow (*Picicorvus columbianus*) and the numerous woodpeckers and jays. The two spermophiles are astonishingly abundant in the lowlands and lower foot-hills, but more and more sparingly distributed up through the Douglas domains,—seldom venturing higher than six or seven thousand feet above the level of the sea. The gray sciurus ranges but little higher than this. The little striped tamias alone is associated with him everywhere. In the lower and middle zones, where they all meet, they are tolerably harmonious—a happy family, though very amusing skirmishes may occasionally be witnessed. Wherever the ancient glaciers have spread forest soil there you find our wee hero, most abundant where depth of soil and genial climate have given rise to a corresponding luxuriance in the trees, but following every kind of growth up the curving moraines to the highest glacial fountains.

Though I cannot of course expect all my readers to sympathize fully in my admiration of this little animal, few, I hope, will think this sketch of his life too long. I cannot begin to tell here how much he has cheered my lonely wanderings during all the years I have been pursuing my studies in these glorious wilds; or how much unmistakable humanity I have found in him. Take this for example: One calm, creamy Indian summer morning, when the nuts were ripe, I was camped in the upper pinewoods of the south fork of the San Joaquin, where the squirrels seemed to be about as plentiful as the ripe burs. They were taking an early breakfast before going to their regular harvest-work. While I was busy with my own breakfast I heard the thudding fall of two or three heavy cones from a Yellow Pine near me. I stole noiselessly forward within about twenty feet of the base of it to observe. In a few moments down came the Douglas. The breakfast-burs he had cut off had rolled on the gently sloping ground into a clump of ceanothus bushes, but he seemed to know exactly where they were, for he found them at once, apparently without searching for them. They were more than twice as heavy as himself, but after turning them into the right position for getting a good hold with his long sickle-teeth he managed to drag them up to the foot of the tree from which he had cut them, moving backward. Then seating himself comfortably, he held them on end, bottom up, and demolished them at his ease. A good deal of nibbling had to be done before he got anything to eat, because the lower scales are barren, but when he had patiently worked his way up to the fertile ones he found two sweet nuts at the base of each, shaped like trimmed hams, and spotted purple like birds' eggs. And notwithstanding these cones were dripping with soft balsam, and covered with prickles, and so strongly put together that a boy would be puzzled to cut them open with a jack-knife, he accomplished his meal with easy dignity and cleanliness, making less effort apparently than a man would in eating soft cookery from a plate.

Breakfast done, I whistled a tune for him before he went to work, curious to see how he would be affected by it. He had not seen me all this while; but the instant I began to whistle he darted up the tree nearest to him, and came out on a small dead limb opposite me, and composed himself to listen. I sang and whistled more than a dozen airs, and as the music changed his eyes sparkled, and

he turned his head quickly from side to side, but
made no other response. Other squirrels, hearing
the strange sounds, came around on all sides, also
chipmunks and birds. One of the birds, a handsome, speckle-breasted thrush, seemed even more
interested than the squirrels. After listening for
awhile on one of the lower dead sprays of a pine,
he came swooping forward within a few feet of my
face, and remained fluttering in the air for half a
minute or so, sustaining himself with whirring
wing-beats, like a humming-bird in front of a flower,
while I could look into his eyes and see his inno-
cent wonder.

By this time my performance must have lasted
nearly half an hour. I sang or whistled "Bonnie
Doon," "Lass o' Gowrie," "O'er the Water to Char-
lie," "Bonnie Woods o' Cragie Lee," etc., all of
which seemed to be listened to with bright interest,
my first Douglas sitting patiently through it all,
with his telling eyes fixed upon me until I ventured to give the "Old Hundredth," when he
screamed his Indian name, Pillillooeet, turned tail,
and darted with ludicrous haste up the tree out of
sight, his voice and actions in the case leaving a
somewhat profane impression, as if he had said,
"I 'll be hanged if you get me to hear anything
so solemn and unpiny." This acted as a signal
for the general dispersal of the whole hairy
tribe, though the birds seemed willing to wait
further developments, music being naturally more
in their line.

What there can be in that grand old church-tune
that is so offensive to birds and squirrels I can't

imagine. A year or two after this High Sierra concert, I was sitting one fine day on a hill in the Coast Range where the common Ground Squirrels were abundant. They were very shy on account of being hunted so much; but after I had been silent and motionless for half an hour or so they began to venture out of their holes and to feed on the seeds of the grasses and thistles around me as if I were no more to be feared than a tree-stump. Then it occurred to me that this was a good opportunity to find out whether they also disliked "Old Hundredth." Therefore I began to whistle as nearly as I could remember the same familiar airs that had pleased the mountaineers of the Sierra. They at once stopped eating, stood erect, and listened patiently until I came to "Old Hundredth," when with ludicrous haste every one of them rushed to their holes and bolted in, their feet twinkling in the air for a moment as they vanished.

No one who makes the acquaintance of our forester will fail to admire him; but he is far too self-reliant and warlike ever to be taken for a darling.

How long the life of a Douglas Squirrel may be, I don't know. The young seem to sprout from knot-holes, perfect from the first, and as enduring as their own trees. It is difficult, indeed, to realize that so condensed a piece of sun-fire should ever become dim or die at all. He is seldom killed by hunters, for he is too small to encourage much of their attention, and when pursued in settled regions becomes excessively shy, and keeps close in the furrows of the highest trunks, many of which are of the same color as himself. Indian boys, however,

16

lie in wait with unbounded patience to shoot them with arrows. In the lower and middle zones a few fall a prey to rattlesnakes. Occasionally he is pursued by hawks and wildcats, etc. But, upon the whole, he dwells safely in the deep bosom of the woods, the most highly favored of all his happy tribe. May his tribe increase!

TRYING THE BOW.

CHAPTER X

A WIND-STORM IN THE FORESTS

THE mountain winds, like the dew and rain, sunshine and snow, are measured and bestowed with love on the forests to develop their strength and beauty. However restricted the scope of other forest influences, that of the winds is universal. The snow bends and trims the upper forests every winter, the lightning strikes a single tree here and there, while avalanches mow down thousands at a swoop as a gardener trims out a bed of flowers. But the winds go to every tree, fingering every leaf and branch and furrowed bole; not one is forgotten; the Mountain Pine towering with outstretched arms on the rugged buttresses of the icy peaks, the lowliest and most retiring tenant of the dells; they seek and find them all, caressing them tenderly, bending them in lusty exercise, stimulating their growth, plucking off a leaf or limb as required, or removing an entire tree or grove, now whispering and cooing through the branches like a sleepy child, now roaring like the ocean; the winds blessing the forests, the forests the winds, with ineffable beauty and harmony as the sure result.

After one has seen pines six feet in diameter bending like grasses before a mountain gale, and

A WIND-STORM IN THE CALIFORNIA FORESTS.
(AFTER A SKETCH BY THE AUTHOR.)

ever and anon some giant falling with a crash that
shakes the hills, it seems astonishing that any, save
the lowest thickset trees, could ever have found a
period sufficiently stormless to establish themselves; or, once established, that they should not,

sooner or later, have been blown down. But when the storm is over, and we behold the same forests tranquil again, towering fresh and unscathed in erect majesty, and consider what centuries of storms have fallen upon them since they were first planted, —hail, to break the tender seedlings; lightning, to scorch and shatter; snow, winds, and avalanches, to crush and overwhelm,—while the manifest result of all this wild storm-culture is the glorious perfection we behold; then faith in Nature's forestry is established, and we cease to deplore the violence of her most destructive gales, or of any other storm-implement whatsoever.

There are two trees in the Sierra forests that are never blown down, so long as they continue in sound health. These are the Juniper and the Dwarf Pine of the summit peaks. Their stiff, crooked roots grip the storm-beaten ledges like eagles' claws, while their lithe, cord-like branches bend round compliantly, offering but slight holds for winds, however violent. The other alpine conifers—the Needle Pine, Mountain Pine, Two-leaved Pine, and Hemlock Spruce—are never thinned out by this agent to any destructive extent, on account of their admirable toughness and the closeness of their growth. In general the same is true of the giants of the lower zones. The kingly Sugar Pine, towering aloft to a height of more than 200 feet, offers a fine mark to storm-winds; but it is not densely foliaged, and its long, horizontal arms swing round compliantly in the blast, like tresses of green, fluent algæ in a brook; while the Silver Firs in most places keep their ranks well together in united strength. The Yellow or Silver Pine is more frequently overturned than any other tree on the Sierra, because its leaves and branches form a larger mass in proportion to its height, while in many places it is planted sparsely, leaving open lanes through which storms may enter with full force. Furthermore, because it is distributed along the lower portion of the range, which was the first to be left bare on the breaking up of the ice-sheet at the close of the glacial winter, the soil it is growing upon has been longer exposed to post-glacial weathering, and consequently is in a more crumbling, decayed condition than the fresher soils farther up the range, and therefore offers a less secure anchorage for the roots.

While exploring the forest zones of Mount Shasta, I discovered the path of a hurricane strewn with thousands of pines of this species. Great and small had been uprooted or wrenched off by sheer force, making a clean gap, like that made by a snow avalanche. But hurricanes capable of doing this class of work are rare in the Sierra, and when we have explored the forests from one extremity of the range to the other, we are compelled to believe that they are the most beautiful on the face of the earth, however we may regard the agents that have made them so.

There is always something deeply exciting, not only in the sounds of winds in the woods, which exert more or less influence over every mind, but in their varied waterlike flow as manifested by the movements of the trees, especially those of the conifers. By no other trees are they rendered so extensively and impressively visible, not even by the lordly tropic palms or tree-ferns responsive to the gentlest breeze. The waving of a forest of the giant Sequoias is indescribably impressive and sublime, but the pines seem to me the best interpreters of winds. They are mighty waving goldenrods, ever in tune, singing and writing wind-music all their long century lives. Little, however, of this noble tree-waving and tree-music will you see or hear in the strictly alpine portion of the forests. The burly Juniper, whose girth sometimes more than equals its height, is about as rigid as the rocks on which it grows. The slender lash-like sprays of the Dwarf Pine stream out in wavering ripples, but the tallest and slenderest are far too unyielding to wave even in the heaviest gales. They only shake in quick, short vibrations. The Hemlock Spruce, however, and the Mountain Pine, and some of the tallest thickets of the Two-leaved species bow in storms with considerable scope and gracefulness. But it is only in the lower and middle zones that the meeting of winds and woods is to be seen in all its grandeur.

One of the most beautiful and exhilarating storms I ever enjoyed in the Sierra occurred in December, 1874, when I happened to be exploring one of the tributary valleys of the Yuba River. The sky and the ground and the trees had been thoroughly rain-washed and were dry again. The day was intensely pure, one of those incomparable bits of California winter, warm and balmy and full of white sparkling sunshine, redolent of all the purest influences of the spring, and at the same time enlivened with one of the most bracing wind-storms conceivable. Instead of camping out, as I usually do, I then chanced to be stopping at the house of a friend. But when the storm began to sound, I lost no time in pushing out into the woods to enjoy it. For on such occasions Nature has always something rare to show us, and the danger to life and limb is hardly greater than one would experience crouching deprecatingly beneath a roof.

It was still early morning when I found myself fairly adrift. Delicious sunshine came pouring over the hills, lighting the tops of the pines, and setting free a steam of summery fragrance that contrasted strangely with the wild tones of the storm. The air was mottled with pine-tassels and bright green plumes, that went flashing past in the sunlight like birds pursued. But there was not the slightest dustiness, nothing less pure than leaves, and ripe pollen, and flecks of withered bracken and moss. I heard trees falling for hours at the rate of one every two or three minutes; some uprooted, partly on account of the loose, water-soaked condition of the ground; others broken straight across, where some weakness caused by fire had determined the spot. The gestures of the various trees made a delightful study. Young Sugar Pines, light and feathery as squirrel-tails, were bowing almost to the ground; while the grand old patriarchs, whose massive boles had been tried in a hundred storms, waved solemnly above them, their long, arching branches streaming fluently on the gale, and every needle thrilling and ringing and shedding off keen lances of light like a diamond. The Douglas Spruces, with long sprays drawn out in level tresses, and needles massed in a gray, shimmering glow, presented a most striking appearance as they stood in bold relief along the hilltops. The madroños in the dells, with their red bark and large glossy leaves tilted every way, reflected the sunshine in throbbing spangles like those one so often sees on the rippled surface of a glacier lake. But the Silver Pines were now the most impressively beautiful of all. Colossal spires 200 feet in height waved like supple golden-rods chanting and bowing low as if in worship, while the whole mass of their long, tremulous foliage was kindled into one continuous blaze of white sun-fire. The force of the gale was such that the most steadfast monarch of them all rocked down to its roots with a motion plainly perceptible when one leaned against it. Nature was holding high festival, and every fiber of the most rigid giants thrilled with glad excitement.

I drifted on through the midst of this passionate music and motion, across many a glen, from ridge to ridge; often halting in the lee of a rock for shelter, or to gaze and listen. Even when the grand anthem had swelled to its highest pitch, I could distinctly hear the varying tones of individual trees,—Spruce, and Fir, and Pine, and leafless Oak,—and even the infinitely gentle rustle of the withered grasses at my feet. Each was expressing itself in its own way,—singing its own song, and making its own peculiar gestures,—manifesting a richness of variety to be found in no other forest I have yet seen. The coniferous woods of Canada, and the Carolinas, and Florida, are made up of trees that resemble one another about as nearly as blades of grass, and grow close together in much the same way. Coniferous trees, in general, seldom possess individual character, such as is manifest among Oaks and Elms. But the California forests are made up of a greater number of distinct species than any other in the world. And in them we find, not only a marked differentiation into special groups, but also a marked individuality in almost every tree, giving rise to storm effects indescribably glorious.

Toward midday, after a long, tingling scramble through copses of hazel and ceanothus, I gained the summit of the highest ridge in the neighborhood; and then it occurred to me that it would be a fine thing to climb one of the trees to obtain a wider outlook and get my ear close to the Æolian music of its topmost needles. But under the circumstances the choice of a tree was a serious matter. One whose instep was not very strong seemed in danger of being blown down, or of being struck by others in case they should fall; another was branchless to a considerable height above the ground, and at the same time too large to be grasped with arms and legs in climbing; while others were not favorably situated for clear views. After cautiously casting about, I made choice of the tallest of a group of Douglas Spruces that were growing close together like a tuft of grass, no one of which seemed likely to fall unless all the rest fell with it. Though comparatively young, they were about 100 feet high, and their lithe,

brushy tops were rocking and swirling in wild ecstasy. Being accustomed to climb trees in making
botanical studies, I experienced no difficulty in
reaching the top of this one, and never before did I
enjoy so noble an exhilaration of motion. The
slender tops fairly flapped and swished in the passionate torrent, bending and swirling backward
and forward, round and round, tracing indescribable combinations of vertical and horizontal curves,
while I clung with muscles firm braced, like a bobolink on a reed.

In its widest sweeps my tree-top described an
arc of from twenty to thirty degrees, but I felt
sure of its elastic temper, having seen others of
the same species still more severely tried—bent
almost to the ground indeed, in heavy snows—without breaking a fiber. I was therefore safe, and free
to take the wind into my pulses and enjoy the excited forest from my superb outlook. The view
from here must be extremely beautiful in any
weather. Now my eye roved over the piny hills
and dales as over fields of waving grain, and felt
the light running in ripples and broad swelling undulations across the valleys from ridge to ridge, as
the shining foliage was stirred by corresponding
waves of air. Oftentimes these waves of reflected
light would break up suddenly into a kind of
beaten foam, and again, after chasing one another
in regular order, they would seem to bend forward
in concentric curves, and disappear on some hillside, like sea-waves on a shelving shore. The
quantity of light reflected from the bent needles
was so great as to make whole groves appear as if

YELLOW PINE AND LIBOCEDRUS.
The two inside trees are Libocedrus, the two outside trees, Yellow Pine.

covered with snow, while the black shadows beneath the trees greatly enhanced the effect of the silvery splendor.

Excepting only the shadows there was nothing somber in all this wild sea of pines. On the contrary, notwithstanding this was the winter season, the colors were remarkably beautiful. The shafts of the pine and libocedrus were brown and purple, and most of the foliage was well tinged with yellow; the laurel groves, with the pale undersides of their leaves turned upward, made masses of gray; and then there was many a dash of chocolate color from clumps of manzanita, and jet of vivid crimson from the bark of the madroños, while the ground on the hillsides, appearing here and there through openings between the groves, displayed masses of pale purple and brown.

The sounds of the storm corresponded gloriously with this wild exuberance of light and motion. The profound bass of the naked branches and boles booming like waterfalls; the quick, tense vibrations of the pine-needles, now rising to a shrill, whistling hiss, now falling to a silky murmur; the rustling of laurel groves in the dells, and the keen metallic click of leaf on leaf — all this was heard in easy analysis when the attention was calmly bent.

The varied gestures of the multitude were seen to fine advantage, so that one could recognize the different species at a distance of several miles by this means alone, as well as by their forms and colors, and the way they reflected the light. All seemed strong and comfortable, as if really enjoy-

ing the storm, while responding to its most enthusiastic greetings. We hear much nowadays
concerning the universal struggle for existence,
but no struggle in the common meaning of the
word was manifest here; no recognition of danger
by any tree; no deprecation; but rather an invin-
cible gladness as remote from exultation as from
fear.

I kept my lofty perch for hours, frequently clos-
ing my eyes to enjoy the music by itself, or to
feast quietly on the delicious fragrance that was
streaming past. The fragrance of the woods was
less marked than that produced during warm rain,
when so many balsamic buds and leaves are
steeped like tea; but, from the chafing of resiny
branches against each other, and the incessant
attrition of myriads of needles, the gale was spiced
to a very tonic degree. And besides the fragrance
from these local sources there were traces of scents
brought from afar. For this wind came first from
the sea, rubbing against its fresh, briny waves,
then distilled through the redwoods, threading rich
ferny gulches, and spreading itself in broad undu-
lating currents over many a flower-enameled ridge
of the coast mountains, then across the golden
plains, up the purple foot-hills, and into these piny
woods with the varied incense gathered by the way.

Winds are advertisements of all they touch,
however much or little we may be able to read
them; telling their wanderings even by their scents
alone. Mariners detect the flowery perfume of
land-winds far at sea, and sea-winds carry the fragrance of dulse and tangle far inland, where it is

quickly recognized, though mingled with the scents
of a thousand land-flowers. As an illustration of
this, I may tell here that I breathed sea-air on
the Firth of Forth, in Scotland, while a boy;
then was taken to Wisconsin, where I remained
nineteen years; then, without in all this time having breathed one breath of the sea, I walked
quietly, alone, from the middle of the Mississippi
Valley to the Gulf of Mexico, on a botanical excursion, and while in Florida, far from the coast, my
attention wholly bent on the splendid tropical
vegetation about me, I suddenly recognized a sea-
breeze, as it came sifting through the palmettos
and blooming vine-tangles, which at once awakened and set free a thousand dormant associations,
and made me a boy again in Scotland, as if all the
intervening years had been annihilated.

Most people like to look at mountain rivers, and
bear them in mind; but few care to look at the
winds, though far more beautiful and sublime, and
though they become at times about as visible as
flowing water. When the north winds in winter
are making upward sweeps over the curving summits of the High Sierra, the fact is sometimes published with flying snow-banners a mile long. Those
portions of the winds thus embodied can scarce be
wholly invisible, even to the darkest imagination.
And when we look around over an agitated forest,
we may see something of the wind that stirs it, by
its effects upon the trees. Yonder it descends in a
rush of water-like ripples, and sweeps over the
bending pines from hill to hill. Nearer, we see
detached plumes and leaves, now speeding by on

level currents, now whirling in eddies, or, escaping over the edges of the whirls, soaring aloft on grand, upswelling domes of air, or tossing on flame-like crests. Smooth, deep currents, cascades, falls, and swirling eddies, sing around every tree and leaf, and over all the varied topography of the region with telling changes of form, like mountain rivers conforming to the features of their channels.

After tracing the Sierra streams from their fountains to the plains, marking where they bloom white in falls, glide in crystal plumes, surge gray and foam-filled in boulder-choked gorges, and slip through the woods in long, tranquil reaches—after thus learning their language and forms in detail, we may at length hear them chanting all together in one grand anthem, and comprehend them all in clear inner vision, covering the range like lace. But even this spectacle is far less sublime and not a whit more substantial than what we may behold of these storm-streams of air in the mountain woods.

We all travel the milky way together, trees and men; but it never occurred to me until this storm-day, while swinging in the wind, that trees are travelers, in the ordinary sense. They make many journeys, not extensive ones, it is true; but our own little journeys, away and back again, are only little more than tree-wavings — many of them not so much.

When the storm began to abate, I dismounted and sauntered down through the calming woods. The storm-tones died away, and, turning toward the east, I beheld the countless hosts of the forests hushed and tranquil, towering above one another on the slopes of the hills like a devout audience. The setting sun filled them with amber light, and seemed to say, while they listened, " My peace I give unto you."

As I gazed on the impressive scene, all the so-called ruin of the storm was forgotten, and never before did these noble woods appear so fresh, so joyous, so immortal.

CHAPTER XI

THE Sierra rivers are flooded every spring by the melting of the snow as regularly as the famous old Nile. They begin to rise in May, and in June high-water mark is reached. But because the melting does not go on rapidly over all the fountains, high and low, simultaneously, and the melted snow is not reinforced at this time of year by rain, the spring floods are seldom very violent or destructive. The thousand falls, however, and the cascades in the cañons are then in full bloom, and sing songs from one end of the range to the other. Of course the snow on the lower tributaries of the rivers is first melted, then that on the higher fountains most exposed to sunshine, and about a month later the cooler, shadowy fountains send down their treasures, thus allowing the main trunk streams nearly six weeks to get their waters hurried through the foot-hills and across the lowlands to the sea. Therefore very violent spring floods are avoided, and will be as long as the shading, restraining forests last. The rivers of the north half of the range are still less subject to sudden floods, because their upper fountains in great part lie protected from the changes of the weather beneath thick folds of lava, just as many of the rivers of Alaska lie beneath folds of ice, coming to the light farther down the range in large springs, while those of the high Sierra lie on the surface of solid granite, exposed to every change of temperature. More than ninety per cent. of the water derived from the snow and ice of Mount Shasta is at once absorbed and drained away beneath the porous lava folds of the mountain, where mumbling and groping in the dark they at length find larger fissures and tunnel-like caves from which they emerge, filtered and cool, in the form of large springs, some of them so large they give birth to rivers that set out on their journeys beneath the sun without any visible intermediate period of childhood. Thus the Shasta River issues from a large lake-like spring in Shasta Valley, and about two thirds of the volume of the McCloud River gushes forth suddenly from the face of a lava bluff in a roaring spring seventy-five yards wide.

These spring rivers of the north are of course shorter than those of the south whose tributaries extend up to the tops of the mountains. Fall River, an important tributary of the Pitt or Upper Sacramento, is only about ten miles long, and is all falls, cascades, and springs from its head to its confluence with the Pitt. Bountiful springs, charmingly embowered, issue from the rocks at one end of it, a snowy fall a hundred and eighty feet high thunders at the other, and a rush of crystal rapids sing and dance between. Of course such streams are but little affected by the weather. Sheltered from evaporation their flow is nearly as full in the autumn as in the time of general spring floods. While those of the high Sierra diminish to less than the hundredth part of their springtime prime, shallowing in autumn to a series of silent pools among the rocks and hollows of their channels, connected by feeble, creeping threads of water, like the sluggish sentences of a tired writer, connected by a drizzle of " ands " and " buts." Strange to say, the greatest floods occur in winter, when one would suppose all the wild waters would be muffled and chained in frost and·snow.   The same long, all-day storms of the so-called Rainy Season in California, that give rain to the lowlands, give dry frosty snow to the mountains. But at rare intervals warm rains and warm winds invade the mountains and push back the snow line from 2000 feet to 8000, or even higher, and then come the big floods.

I was usually driven down out of the High Sierra about the end of November, but the winter of 1874 and 1875 was so warm and calm that I was tempted to seek general views of the geology and topography of the basin of Feather River in January.   And I had just completed a hasty survey of the region, and made my way down to winter quarters, when one of the grandest flood-storms that I ever saw broke on the mountains.   I was then in the edge of the main forest belt at a small foot-hill town called Knoxville, on the divide between the waters of the Feather and Yuba rivers. The cause of this notable flood was simply a sudden and copious fall of warm wind and rain on the basins of these rivers at a time when they contained a considerable quantity of snow. The rain was so heavy and long-sustained that it was, of itself, sufficient to make a good wild flood, while the snow which the warm wind and rain melted on the upper and middle regions of the basins was sufficient to make another flood equal to that of the rain. Now these two distinct harvests of flood waters were gathered simultaneously and poured out on the plain in one magnificent avalanche. The basins of the Yuba and Feather, like many others of the Sierra, are admirably adapted to the growth of floods of this kind. Their many tributaries radiate far and wide, comprehending extensive areas, and the tributaries are steeply inclined, while the trunks are comparatively level. While the flood-storm was in progress the thermometer at Knoxville ranged between 44° and 50°; and when warm wind and warm rain fall simultaneously on snow contained in basins like these, both the rain and that portion of the snow which the rain and wind melt are at first sponged up and held back until the combined mass becomes sludge, which at length, suddenly dissolving, slips and descends all together to the trunk channel; and since the deeper the stream the faster it flows, the flooded portion of the current above overtakes the slower foot-hill portion below it, and all sweeping forward together with a high, overcurling front, debouches on the open plain with a violence and suddenness that at first seem wholly unaccountable. The destructiveness of the lower portion of this particular flood was somewhat augmented by mining gravel in the river channels, and by levees which gave way after having at first restrained and held back the accumulating waters. These exaggerating conditions did not, however, greatly influence the general result, the main effect having been caused by the rare combination of flood factors indicated above. It is a pity that but few people meet and enjoy storms so noble as this in their homes in the mountains, for, spending themselves in the open levels of the plains, they are likely to be remembered more by the bridges and houses they carry away than by their beauty or the thousand blessings they bring to the fields and gardens of Nature.

On the morning of the flood, January 19th, all the Feather and Yuba landscapes were covered with running water, muddy torrents filled every gulch and ravine, and the sky was thick with rain. The pines had long been sleeping in sunshine; they were now awake, roaring and waving with the beating storm, and the winds sweeping along the curves of hill and dale, streaming through the woods, surging and gurgling on the tops of rocky ridges, made the wildest of wild storm melody.

It was easy to see that only a small part of the rain reached the ground in the form of drops. Most of it was thrashed into dusty spray like that into which small waterfalls are divided when they dash on shelving rocks. Never have I seen water coming from the sky in denser or more passionate streams. The wind chased the spray forward in choking drifts, and compelled me again and again to seek shelter in the dell copses and back of large trees to rest and catch my breath. Wherever I went, on ridges or in hollows, enthusiastic water still flashed and gurgled about my ankles, recalling a wild winter flood in Yosemite when a hundred waterfalls came booming and chanting together and filled the grand valley with a sea-like roar.

After drifting an hour or two in the lower woods, I set out for the summit of a hill 900 feet high, with a view to getting as near the heart of the storm as possible. In order to reach it I had to cross Dry Creek, a tributary of the Yuba that goes crawling along the base of the hill on the northwest. It was now a booming river as large as the Tuolumne at ordinary stages, its current brown with mining-mud washed down from many a " claim," and mottled with sluice-boxes, fence-rails, and logs that had long lain above its reach. A slim foot-bridge stretched across it, now scarcely above the swollen current. Here I was glad to linger, gazing and listening, while the storm was in its richest mood — the gray rain-flood above, the brown river-flood beneath. The language of the river was scarcely less enchanting than that of the wind and rain; the sublime overboom of the main bouncing, exulting current, the swash and gurgle of the eddies, the keen dash and clash of heavy waves breaking against rocks, and the smooth, downy hush of shallow currents feeling their way through the willow thickets of the margin. And amid all this varied throng of sounds I heard the smothered bumping and rumbling of boulders on the bottom as they were shoving and rolling forward against one another in a wild rush, after having lain still for probably 100 years or more.

The glad creek rose high above its banks and wandered from its channel out over many a briery sand-flat and meadow. Alders and willows waist-deep were bearing up against the current with nervous trembling gestures, as if afraid of being carried away, while supple branches bending confidingly, dipped lightly and rose again, as if stroking the wild waters in play. Leaving the bridge and passing on through the storm-thrashed woods, all the ground seemed to be moving. Pine-tassels, flakes of bark, soil, leaves, and broken branches were being swept forward, and many a rock-fragment, weathered from exposed ledges, was now receiving its first rounding and polishing in the wild streams of the storm. On they rushed through every gulch and hollow, leaping, gliding, working with a will, and rejoicing like living creatures.

Nor was the flood confined to the ground. Every tree had a water system of its own spreading far and wide like miniature Amazons and Mississippis.

Toward midday, cloud, wind, and rain reached their highest development. The storm was in full bloom, and formed, from my commanding outlook on the hilltop, one of the most glorious views I ever beheld. As far as the eye could reach, above, beneath, around, wind-driven rain filled the air like one vast waterfall. Detached clouds swept imposingly up the valley, as if they were endowed with independent motion and had special work to do in replenishing the mountain wells, now rising above the pine-tops, now descending into their midst, fondling their arrowy spires and soothing every branch and leaf with gentleness in the midst of all the savage sound and motion. Others keeping near the ground glided behind separate groves, and brought them forward into relief with admirable distinctness; or, passing in front, eclipsed whole groves in succession, pine after pine melting in their gray fringes and bursting forth again seemingly clearer than before.

The forms of storms are in great part measured, and controlled by the topography of the regions where they rise and over which they pass. When, therefore, we attempt to study them from the valleys, or from gaps and openings of the forest, we are confounded by a multitude of separate and apparently antagonistic impressions. The bottom of the storm is broken up into innumerable waves and currents that surge against the hillsides like sea-waves against a shore, and these, reacting on the nether surface of the storm, erode immense cavernous hollows and cañons, and sweep forward the resulting detritus in long trains, like the moraines of glaciers. But, as we ascend, these partial, confusing effects disappear and the phenomena are beheld united and harmonious.

The longer I gazed into the storm, the more plainly visible it became. The drifting cloud detritus gave it a kind of visible body, which explained many perplexing phenomena, and published its movements in plain terms, while the texture of the falling mass of rain rounded it out and rendered it more complete. Because raindrops differ in size they fall at different velocities and overtake and clash against one another, producing mist and spray. They also, of course, yield unequal compliance to the force of the wind, which gives rise to a still greater degree of interference, and passionate gusts sweep off clouds of spray from the groves like that torn from wave-tops in a gale. All these factors of irregularity in density, color, and texture of the general rain mass tend to make it the more appreciable and telling. It is then seen as one grand flood rushing over bank and brae, bending the pines like weeds, curving this way and that, whirling in huge eddies in hollows and dells, while the main current pours grandly over all, like ocean currents over the landscapes that lie hidden at the bottom of the sea.

I watched the gestures of the pines while the storm was at its height, and it was easy to see that they were not distressed. Several large Sugar Pines stood near the thicket in which I was sheltered, bowing solemnly and tossing their long arms as if interpreting the very words of the storm while accepting its wildest onsets with passionate exhilaration. The lions were feeding. Those who have observed sunflowers feasting on sunshine during the golden days of Indian summer know that none of their gestures express thankfulness. Their celestial food is too heartily given, too heartily taken to leave room for thanks. The pines were evidently accepting the benefactions of the storm in the same whole-souled manner; and when I looked down among the budding hazels, and still lower to the young violets and fern-tufts on the rocks, I noticed the same divine methods of giving and taking, and the same exquisite adaptations of what seems an outbreak of violent and uncontrollable force to the purposes of beautiful and delicate life. Calms like sleep come upon landscapes, just as they do on people and trees, and storms awaken them in the same way. In the dry midsummer of the lower portion of the range the withered hills and valleys seem to lie as empty and expressionless as dead shells on a shore. Even the highest mountains may be found occasionally dull and uncommunicative as if in some way they had lost countenance and shrunk to less than half their real stature. But when the lightnings crash and echo in the cañons, and the clouds come down wreathing and crowning their bald snowy heads, every feature beams with expression and they rise again in all their imposing majesty.

Storms are fine speakers, and tell all they know, but their voices of lightning, torrent, and rushing wind are much less numerous than the nameless still, small voices too low for human ears; and because we are poor listeners we fail to catch much that is fairly within reach. Our best rains are heard mostly on roofs, and winds in chimneys; and when by choice or compulsion we are pushed into the heart of a storm, the confusion made by cumbersome equipments and nervous haste and mean fear, prevent our hearing any other than the loudest expressions. Yet we may draw enjoyment from storm sounds that are beyond hearing, and storm movements we cannot see. The sublime whirl of planets around their suns is as silent as raindrops oozing in the dark among the roots of plants. In this great storm, as in every other, there were tones and gestures inexpressibly gentle manifested in the midst of what is called violence and fury, but easily recognized by all who look and listen for them. The rain brought out the colors of the woods with delightful freshness, the rich brown of the bark of the trees and the fallen burs and leaves and dead ferns; the grays of rocks and lichens; the light purple of swelling buds, and the warm yellow greens of the libocedrus and mosses. The air was steaming with delightful fragrance, not rising and wafting past in separate masses, but diffused through all the atmosphere. Pine woods are always fragrant, but most so in spring when the young tassels are opening and in warm weather when the various gums and balsams are softened by the sun. The wind was now chafing their innumerable needles and the warm rain was steeping them. Monardella grows here in large beds in the openings, and there is plenty of laurel in dells and manzanita on the hillsides, and the rosy, fragrant chamœbatia carpets the ground almost everywhere. These, with the gums and balsams of the woods, form the main local fragrance-fountains of the storm. The ascending clouds of aroma wind-rolled and rain-washed became pure like light and traveled with the wind as part of it. Toward the middle of the afternoon the main flood cloud lifted along its western border revealing a beautiful section of the Sacramento Valley some twenty or thirty miles away, brilliantly sun-lighted and glistering with rain-sheets as if paved with silver. Soon afterward a jagged bluff-like cloud with a sheer face appeared over the valley of the Yuba, dark-colored and roughened with numerous

furrows like some huge lava-table. The blue Coast
Range was seen stretching along the sky like a
beveled wall, and the somber, craggy Marysville
Buttes rose impressively out of the flooded plain
like islands out of the sea. Then the rain began to
abate and I sauntered down through the dripping
bushes reveling in the universal vigor and freshness that inspired all the life about me. How
clean and unworn and immortal the woods seemed
to be!—the lofty cedars in full bloom laden with
golden pollen and their washed plumes shining;
the pines rocking gently and settling back into
rest, and the evening sunbeams spangling on the
broad leaves of the madroños, their tracery of
yellow boughs relieved against dusky thickets of
Chestnut Oak; liverworts, lycopodiums, ferns were
exulting in glorious revival, and every moss that
had ever lived seemed to be coming crowding back
from the dead to clothe each trunk and stone in
living green. The steaming ground seemed fairly
to throb and tingle with life; smilax, fritillaria,
saxifrage, and young violets were pushing up as if
already conscious of the summer glory, and innumerable green and yellow buds were peeping
and smiling everywhere.

As for the birds and squirrels, not a wing or tail
of them was to be seen while the storm was blowing. Squirrels dislike wet weather more than cats
do; therefore they were at home rocking in their dry
nests. The birds were hiding in the dells out of
the wind, some of the strongest of them pecking
at acorns and manzanita berries, but most were
perched on low twigs, their breast feathers puffed

out and keeping one another company through the hard time as best they could.

When I arrived at the village about sundown, the good people bestirred themselves, pitying my bedraggled condition as if I were some benumbed castaway snatched from the sea, while I, in turn, warm with excitement and reeking like the ground, pitied them for being dry and defrauded of all the glory that Nature had spread round about them that day.

CHAPTER XII

SIERRA THUNDER-STORMS

THE weather of spring and summer in the middle region of the Sierra is usually well flecked with rains and light dustings of snow, most of which are far too obviously joyful and life-giving to be regarded as storms; and in the picturesque beauty and clearness of outlines of their clouds they offer striking contrasts to those boundless, all-embracing cloud-mantles of the storms of winter. The smallest and most perfectly individualized specimens present a richly modeled cumulous cloud rising above the dark woods, about 11 A. M., swelling with a visible motion straight up into the calm, sunny sky to a height of 12,000 to 14,000 feet above the sea, its white, pearly bosses relieved by gray and pale purple shadows in the hollows, and showing outlines as keenly defined as those of the glacier-polished domes. In less than an hour it attains full development and stands poised in the blazing sunshine like some colossal mountain, as beautiful in form and finish as if it were to become a permanent addition to the landscape. Presently a thunderbolt crashes through the crisp air, ringing like steel on steel, sharp and clear, its startling detonation breaking into a spray of echoes against

the cliffs and cañon walls. Then down comes a
cataract of rain. The big drops sift through the
pine-needles, plash and patter on the granite pavements, and pour down the sides of ridges and
domes in a network of gray, bubbling rills. In a
few minutes the cloud withers to a mesh of dim
filaments and disappears, leaving the sky perfectly
clear and bright, every dust-particle wiped and
washed out of it. Everything is refreshed and in-
vigorated, a steam of fragrance rises, and the
storm is finished — one cloud, one lightning-stroke,
and one dash of rain. This is the Sierra midsummer thunder-storm reduced to its lowest terms.
But some of them attain much larger proportions,
and assume a grandeur and energy of expression
hardly surpassed by those bred in the depths of
winter, producing those sudden floods called
"cloud-bursts," which are local, and to a considerable extent periodical, for they appear nearly every
day about the same time for weeks, usually about
eleven o'clock, and lasting from five minutes to an
hour or two. One soon becomes so accustomed to
see them that the noon sky seems empty and abandoned without them, as if Nature were forgetting
something. When the glorious pearl and alabaster
clouds of these noonday storms are being built I
never give attention to anything else. No mountain or mountain-range, however divinely clothed
with light, has a more enduring charm than
those fleeting mountains of the sky — floating fountains bearing water for every well, the angels of the
streams and lakes; brooding in the deep azure, or
sweeping softly along the ground over ridge and

BRIDAL VEIL FALLS, YOSEMITE VALLEY.

dome, over meadow, over forest, over garden and grove; lingering with cooling shadows, refreshing every flower, and soothing rugged rock-brows with a gentleness of touch and gesture wholly divine.

The most beautiful and imposing of the summer storms rise just above the upper edge of the Silver Fir zone, and all are so beautiful that it is not easy to choose any one for particular description. The one that I remember best fell on the mountains near Yosemite Valley, July 19, 1869, while I was encamped in the Silver Fir woods. A range of bossy cumuli took possession of the sky, huge domes and peaks rising one beyond another with deep cañons between them, bending this way and that in long curves and reaches, interrupted here and there with white upboiling masses that looked like the spray of waterfalls. Zigzag lances of lightning followed each other in quick succession, and the thunder was so gloriously loud and massive it seemed as if surely an entire mountain was being shattered at every stroke. Only the trees were touched, however, so far as I could see,— a few firs 200 feet high, perhaps, and five to six feet in diameter, were split into long rails and slivers from top to bottom and scattered to all points of the compass. Then came the rain in a hearty flood, covering the ground and making it shine with a continuous sheet of water that, like a transparent film or skin, fitted closely down over all the rugged anatomy of the landscape.

It is not long, geologically speaking, since the first raindrop fell on the present landscapes of the Sierra; and in the few tens of thousands of years

18

of stormy cultivation they have been blest with, how beautiful they have become! The first rains fell on raw, crumbling moraines and rocks without a plant. Now scarcely a drop can fail to find a beautiful mark: on the tops of the peaks, on the smooth glacier pavements, on the curves of the domes, on moraines full of crystals, on the thousand forms of yosemitic sculpture with their tender beauty of balmy, flowery vegetation, laving, plashing, glinting, pattering; some falling softly on meadows, creeping out of sight, seeking and finding every thirsty rootlet, some through the spires of the woods, sifting in dust through the needles, and whispering good cheer to each of them; some falling with blunt tapping sounds, drumming on the broad leaves of veratrum, cypripedium, saxifrage; some falling straight into fragrant corollas, kissing the lips of lilies, glinting on the sides of crystals, on shining grains of gold; some falling into the fountains of snow to swell their well-saved stores; some into the lakes and rivers, patting the smooth glassy levels, making dimples and bells and spray, washing the mountain windows, washing the wandering winds; some plashing into the heart of snowy falls and cascades as if eager to join in the dance and the song and beat the foam yet finer. Good work and happy work for the merry mountain raindrops, each one of them a brave fall in itself, rushing from the cliffs and hollows of the clouds into the cliffs and hollows of the mountains; away from the thunder of the sky into the thunder of the roaring rivers. And how far they have to go, and how many cups to fill—cassiope-cups, holding half a drop, and lake basins between the hills, each replenished with equal care — every drop God's messenger sent on its way with glorious pomp and display of power— silvery new-born stars with lake and river, mountain and valley—all that the landscape holds—reflected in their crystal depths.

CHAPTER XIII

THE WATER-OUZEL

THE waterfalls of the Sierra are frequented by only one bird,—the Ouzel or Water Thrush (*Cinclus Mexicanus*, Sw.). He is a singularly joyous and lovable little fellow, about the size of a robin, clad in a plain waterproof suit of bluish gray, with a tinge of chocolate on the head and shoulders. In form he is about as smoothly plump and compact as a pebble that has been whirled in a pot-hole, the flowing contour of his body being interrupted only by his strong feet and bill, the crisp wing-tips, and the up-slanted wren-like tail.

Among all the countless waterfalls I have met in the course of ten years' exploration in the Sierra, whether among the icy peaks, or warm foot-hills, or in the profound yosemitic cañons of the middle region, not one was found without its Ouzel. No cañon is too cold for this little bird, none too lonely, provided it be rich in falling water. Find a fall, or cascade, or rushing rapid, anywhere upon a clear stream, and there you will surely find its complementary Ouzel, flitting about in the spray, diving in foaming eddies, whirling like a leaf among beaten foam-bells; ever vigorous and enthusiastic, yet self-contained, and neither seeking nor shunning your company.

WATER-OUZEL DIVING AND FEEDING.

If disturbed while dipping about in the margin shallows, he either sets off with a rapid whir to some other feeding-ground up or down the stream, or alights on some half-submerged rock or snag out in the current, and immediately begins to nod and courtesy like a wren, turning his head from side to side with many other odd dainty movements that never fail to fix the attention of the observer.

He is the mountain streams' own darling, the humming-bird of blooming waters, loving rocky ripple-slopes and sheets of foam as a bee loves flowers, as a lark loves sunshine and meadows. Among all the mountain birds, none has cheered me so much in my lonely wanderings,—none so unfailingly. For both in winter and summer he sings, sweetly, cheerily, independent alike of sunshine and of love, requiring no other inspiration than the stream on which he dwells. While water sings, so must he, in heat or cold, calm or storm, ever attuning his voice in sure accord; low in the drought of summer and the drought of winter, but never silent.

During the golden days of Indian summer, after most of the snow has been melted, and the mountain streams have become feeble,—a succession of silent pools, linked together by shallow, transparent currents and strips of silvery lacework,—then the song of the Ouzel is at its lowest ebb. But as soon as the winter clouds have bloomed, and the mountain treasuries are once more replenished with snow, the voices of the streams and ouzels increase in strength and richness until the flood season of early summer. Then the torrents chant their noblest anthems, and then is the flood-time of our songster's melody. As for weather, dark days and sun days are the same to him. The voices of most song-birds, however joyous, suffer a long winter eclipse; but the Ouzel sings on through all the seasons and every kind of storm. Indeed no storm can be more violent than those of the waterfalls in the midst of which he delights to dwell. However dark and boisterous the weather, snowing, blowing, or cloudy, all the same he sings, and with never a note of sadness. No need of spring sunshine to thaw *his* song, for it never freezes. Never shall you hear anything wintry from *his* warm breast; no pinched cheeping, no wavering notes between sorrow and joy; his mellow, fluty voice is ever tuned to downright gladness, as free from dejection as cock-crowing.

It is pitiful to see wee frost-pinched sparrows on cold mornings in the mountain groves shaking the snow from their feathers, and hopping about as if anxious to be cheery, then hastening back to their hidings out of the wind, puffing out their breast-feathers over their toes, and subsiding among the leaves, cold and breakfastless, while the snow continues to fall, and there is no sign of clearing. But the Ouzel never calls forth a single touch of pity; not because he is strong to endure, but rather because he seems to live a charmed life beyond the reach of every influence that makes endurance necessary.

One wild winter morning, when Yosemite Valley was swept its length from west to east by a cordial snow-storm, I sallied forth to see what I might learn and enjoy. A sort of gray, gloaming-like darkness filled the valley, the huge walls were out of sight, all ordinary sounds were smothered, and even the loudest booming of the falls was at times buried beneath the roar of the heavy-laden blast. The loose snow was already over five feet deep on

the meadows, making extended walks impossible
without the aid of snow-shoes. I found no great
difficulty, however, in making my way to a certain
ripple on the river where one of my ouzels lived.
He was at home, busily gleaning his breakfast
among the pebbles of a shallow portion of the
margin, apparently unaware of anything extraordinary in the weather. Presently he flew out to a
stone against which the icy current was beating,
and turning his back to the wind, sang as delightfully as a lark in springtime.

After spending an hour or two with my favorite,
I made my way across the valley, boring and wallowing through the drifts, to learn as definitely as
possible how the other birds were spending their
time. The Yosemite birds are easily found during
the winter because all of them excepting the Ouzel
are restricted to the sunny north side of the valley,
the south side being constantly eclipsed by the
great frosty shadow of the wall. And because the
Indian Cañon groves, from their peculiar exposure,
are the warmest, the birds congregate there, more
especially in severe weather.

I found most of the robins cowering on the lee
side of the larger branches where the snow could
not fall upon them, while two or three of the more
enterprising were making desperate efforts to reach
the mistletoe berries by clinging nervously to the
under side of the snow-crowned masses, back
downward, like woodpeckers. Every now and
then they would dislodge some of the loose fringes
of the snow-crown, which would come sifting down
on them and send them screaming back to camp,

where they would subside among their companions with a shiver, muttering in low, querulous chatter like hungry children.

Some of the sparrows were busy at the feet of the larger trees gleaning seeds and benumbed insects, joined now and then by a robin weary of his unsuccessful attempts upon the snow-covered berries. The brave woodpeckers were clinging to the snowless sides of the larger boles and overarching branches of the camp trees, making short flights from side to side of the grove, pecking now and then at the acorns they had stored in the bark, and chattering aimlessly as if unable to keep still, yet evidently putting in the time in a very dull way, like storm-bound travelers at a country tavern. The hardy nut-hatches were threading the open furrows of the trunks in their usual industrious manner, and uttering their quaint notes, evidently less distressed than their neighbors. The Steller jays were of course making more noisy stir than all the other birds combined; ever coming and going with loud bluster, screaming as if each had a lump of melting sludge in his throat, and taking good care to improve the favorable opportunity afforded by the storm to steal from the acorn stores of the woodpeckers. I also noticed one solitary gray eagle braving the storm on the top of a tall pine-stump just outside the main grove. He was standing bolt upright with his back to the wind, a tuft of snow piled on his square shoulders, a monument of passive endurance. Thus every snow-bound bird seemed more or less uncomfortable if not in positive distress.

The storm was reflected in every gesture, and not one cheerful note, not to say song, came from a single bill; their cowering, joyless endurance offering a striking contrast to the spontaneous, irrepressible gladness of the Ouzel, who could no more help exhaling sweet song than a rose sweet fragrance. He *must* sing though the heavens fall. I remember noticing the distress of a pair of robins during the violent earthquake of the year 1872, when the pines of the Valley, with strange movements, flapped and waved their branches, and beetling rock-brows came thundering down to the meadows in tremendous avalanches. It did not occur to me in the midst of the excitement of other observations to look for the ouzels, but I doubt not they were singing straight on through it all, regarding the terrible rock-thunder as fearlessly as they do the booming of the waterfalls.

What may be regarded as the separate songs of the Ouzel are exceedingly difficult of description, because they are so variable and at the same time so confluent. Though I have been acquainted with my favorite ten years, and during most of this time have heard him sing nearly every day, I still detect notes and strains that seem new to me. Nearly all of his music is sweet and tender, lapsing from his round breast like water over the smooth lip of a pool, then breaking farther on into a sparkling foam of melodious notes, which glow with subdued enthusiasm, yet without expressing much of the strong, gushing ecstasy of the bobolink or skylark.

The more striking strains are perfect arabesques of melody, composed of a few full, round, mellow notes, embroidered with delicate trills which fade and melt in long slender cadences. In a general way his music is that of the streams refined and spiritualized. The deep booming notes of the falls are in it, the trills of rapids, the gurgling of margin eddies, the low whispering of level reaches, and the sweet tinkle of separate drops oozing from the ends of mosses and falling into tranquil pools.

The Ouzel never sings in chorus with other birds, nor with his kind, but only with the streams. And like flowers that bloom beneath the surface of the ground, some of our favorite's best song-blossoms never rise above the surface of the heavier music of the water. I have often observed him singing in the midst of beaten spray, his music completely buried beneath the water's roar; yet I knew he was surely singing by his gestures and the movements of his bill.

His food, as far as I have noticed, consists of all kinds of water insects, which in summer are chiefly procured along shallow margins. Here he wades about ducking his head under water and deftly turning over pebbles and fallen leaves with his bill, seldom choosing to go into deep water where he has to use his wings in diving.

He seems to be especially fond of the larvæ of mosquitos, found in abundance attached to the bottom of smooth rock channels where the current is shallow. When feeding in such places he wades up-stream, and often while his head is under water the swift current is deflected upward along the glossy curves of his neck and shoulders, in the form of a clear, crystalline shell, which fairly incloses him like a bell-glass, the shell being broken and re-formed as he lifts and dips his head; while ever and anon he sidles out to where the too powerful current carries him off his feet; then he dexterously rises on the wing and goes gleaning again in shallower places.

But during the winter, when the stream-banks are embossed in snow, and the streams themselves are chilled nearly to the freezing-point, so that the snow falling into them in stormy weather is not wholly dissolved, but forms a thin, blue sludge, thus rendering the current opaque — then he seeks the deeper portions of the main rivers, where he may dive to clear water beneath the sludge. Or he repairs to some open lake or millpond, at the bottom of which he feeds in safety.

When thus compelled to betake himself to a lake, he does not plunge into it at once like a duck, but always alights in the first place upon some rock or fallen pine along the shore. Then flying out thirty or forty yards, more or less, according to the character of the bottom, he alights with a dainty glint on the surface, swims about, looks down, finally makes up his mind, and disappears with a sharp stroke of his wings. After feeding for two or three minutes he suddenly reappears, showers the water from his wings with one vigorous shake, and rises abruptly into the air as if pushed up from beneath, comes back to his perch, sings a few minutes, and goes out to dive again; thus coming and going, singing and diving at the same place for hours.

The Ouzel is usually found singly; rarely in

pairs, excepting during the
breeding season, and *very* rarely
in threes or fours.   I once observed three thus spending a
winter morning in company,
upon a small glacier lake, on
the Upper Merced,
about 7500 feet above
the level of the sea.
A storm had occurred
during the night,
but the morning sun
shone unclouded, and
the shadowy lake,
gleaming darkly in
its setting of fresh
snow, lay smooth
and motionless as a
mirror.   My camp
chanced to be within
a few feet of the

ONE OF THE LATE-SUMMER FEEDING-
GROUNDS OF THE OUZEL.

water's edge, opposite a fallen pine, some of the
branches of which leaned out over the lake.   Here

my three dearly welcome visitors took up their station, and at once began to embroider the frosty air with their delicious melody, doubly delightful to me that particular morning, as I had been somewhat apprehensive of danger in breaking my way down through the snow-choked cañons to the lowlands.

The portion of the lake bottom selected for a feeding-ground lies at a depth of fifteen or twenty feet below the surface, and is covered with a short growth of algæ and other aquatic plants,—facts I had previously determined while sailing over it on a raft. After alighting on the glassy surface, they occasionally indulged in a little play, chasing one another round about in small circles; then all three would suddenly dive together, and then come ashore and sing.

The Ouzel seldom swims more than a few yards on the surface, for, not being web-footed, he makes rather slow progress, but by means of his strong, crisp wings he swims, or rather flies, with celerity under the surface, often to considerable distances. But it is in withstanding the force of heavy rapids that his strength of wing in this respect is most strikingly manifested. The following may be regarded as a fair illustration of his power of sub-aquatic flight. One stormy morning in winter when the Merced River was blue and green with unmelted snow, I observed one of my ouzels perched on a snag out in the midst of a swift-rushing rapid, singing cheerily, as if everything was just to his mind; and while I stood on the bank admiring him, he suddenly plunged into the sludgy current, leaving his song abruptly broken off. After feeding a minute or two at the bottom, and when one would suppose that he must inevitably be swept far down-stream, he emerged just where he went down, alighted on the same snag, showered the water-beads from his feathers, and continued his unfinished song, seemingly in tranquil ease as if it had suffered no interruption.

The Ouzel alone of all birds dares to enter a white torrent. And though strictly terrestrial in structure, no other is so inseparably related to water, not even the duck, or the bold ocean alba-

tross, or the stormy-petrel. For ducks go ashore as soon as they finish feeding in undisturbed places, and very often make long flights overland from lake to lake or field to field.

OUZEL ENTERING A WHITE CURRENT.

The same is true of most other aquatic birds. But the Ouzel, born on the brink of a stream, or on a snag or boulder in the midst of it, seldom leaves

it for a single moment. For, notwithstanding he
is often on the wing, he never flies overland, but
whirs with rapid, quail-like beat above the stream,
tracing all its windings. Even when the stream is
quite small, say from five to ten feet wide, he seldom shortens his flight by crossing a bend, however abrupt it may be; and even when disturbed
by meeting some one on the bank, he prefers to
fly over one's head, to dodging out over the ground.
When, therefore, his flight along a crooked stream
is viewed endwise, it appears most strikingly wavered—a description on the air of every curve with
lightning-like rapidity.

The vertical curves and angles of the most precipitous torrents he traces with the same rigid
fidelity, swooping down the inclines of cascades,
dropping sheer over dizzy falls amid the spray,
and ascending with the same fearlessness and ease,
seldom seeking to lessen the steepness of the acclivity by beginning to ascend before reaching the
base of the fall. No matter though it may be several hundred feet in height he holds straight on, as
if about to dash headlong into the throng of booming rockets, then darts abruptly upward, and, after
alighting at the top of the precipice to rest a
moment, proceeds to feed and sing. His flight is
solid and impetuous, without any intermission of
wing-beats,— one homogeneous buzz like that of a
laden bee on its way home. And while thus buzzing
freely from fall to fall, he is frequently heard giving
utterance to a long outdrawn train of unmodulated
notes, in no way connected with his song, but corresponding closely with his flight in sustained vigor.

Were the flights of all the ouzels in the Sierra traced on a chart, they would indicate the direction of the flow of the entire system of ancient glaciers, from about the period of the breaking up of the ice-sheet until near the close of the glacial winter; because the streams which the ouzels so rigidly follow are, with the unimportant exceptions of a few side tributaries, all flowing in channels eroded for them out of the solid flank of the range by the vanished glaciers,—the streams tracing the ancient glaciers, the ouzels tracing the streams. Nor do we find so complete compliance to glacial conditions in the life of any other mountain bird, or animal of any kind. Bears frequently accept the pathways laid down by glaciers as the easiest to travel; but they often leave them and cross over from cañon to cañon. So also, most of the birds trace the moraines to some extent, because the forests are growing on them. But they wander far, crossing the cañons from grove to grove, and draw exceedingly angular and complicated courses.

The Ouzel's nest is one of the most extraordinary pieces of bird architecture I ever saw, odd and novel in design, perfectly fresh and beautiful, and in every way worthy of the genius of the little builder. It is about a foot in diameter, round and bossy in outline, with a neatly arched opening near the bottom, somewhat like an old-fashioned brick oven, or Hottentot's hut. It is built almost exclusively of green and yellow mosses, chiefly the beautiful fronded hypnum that covers the rocks and old drift-logs in the vicinity of waterfalls. These are deftly interwoven, and felted together

19

into a charming little hut; and so situated that many of the outer mosses continue to flourish as if they had not been plucked. A few fine, silky-stemmed grasses are occasionally found interwoven with the mosses, but, with the exception of a thin layer lining the floor, their presence seems accidental, as they are of a species found growing with the mosses and are probably plucked with them. The site chosen for this curious mansion is usually some little rock-shelf within reach of the lighter particles of the spray of a waterfall, so that its walls are kept green and growing, at least during the time of high water.

No harsh lines are presented by any portion of the nest as seen in place, but when removed from its shelf, the back and bottom, and sometimes a portion of the top, is found quite sharply angular, because it is made to conform to the surface of the rock upon which and against which it is built, the little architect always taking advantage of slight crevices and protuberances that may chance to offer, to render his structure stable by means of a kind of gripping and dovetailing.

In choosing a building-spot, concealment does not seem to be taken into consideration; yet notwithstanding the nest is large and guilelessly exposed to view, it is far from being easily detected, chiefly because it swells forward like any other bulging moss-cushion growing naturally in such situations. This is more especially the case where the nest is kept fresh by being well sprinkled. Sometimes these romantic little huts have their beauty enhanced by rock-ferns and grasses that spring up around the mossy walls, or in front of the door-sill, dripping with crystal beads.

Furthermore, at certain hours of the day, when the sunshine is poured down at the required angle, the whole mass of the spray enveloping the fairy establishment is brilliantly irised; and it is through so glorious a rainbow atmosphere as this that some of our blessed ouzels obtain their first peep at the world.

Ouzels seem so completely part and parcel of the streams they inhabit, they scarce suggest any other origin than the streams themselves; and one might almost be pardoned in fancying they come direct from the living waters, like flowers from the ground. At least, from whatever cause, it never occurred to me to look for their nests until more than a year after I had made the acquaintance of the birds themselves, although I found one the very day on which I began the search. In making my way from Yosemite to the glaciers at the heads of the Merced and Tuolumne rivers, I camped in a particularly wild and romantic portion of the Nevada cañon where in previous excursions I had never failed to enjoy the company of my favorites, who were attracted here, no doubt, by the safe nesting-places in the shelving rocks, and by the abundance of food and falling water. The river, for miles above and below, consists of a succession of small falls from ten to sixty feet in height, connected by flat, plume-like cascades that go flashing from fall to fall, free and almost channelless, over waving folds of glacier-polished granite.

On the south side of one of the falls, that portion of the precipice which is bathed by the spray presents a series of little shelves and tablets caused by the development of planes of cleavage in the granite, and by the consequent fall of masses through the action of the water. "Now here," said I, " of all places, is the most charming spot for an Ouzel's nest." Then carefully scanning the fretted face of the precipice through the spray, I at length noticed a yellowish moss-cushion, growing on the edge of a level tablet within five or six feet of the outer folds of the fall. But apart from the fact of its being situated where one acquainted with the lives of ouzels would fancy an Ouzel's nest ought to be, there was nothing in its appearance visible at first sight, to distinguish it from other bosses of rock-moss similarly situated with reference to perennial spray; and it was not until I had scrutinized it again and again, and had removed my shoes and stockings and crept along the face of the rock within eight or ten feet of it, that I could decide certainly whether it was a nest or a natural growth.

In these moss huts three or four eggs are laid, white like foam-bubbles; and well may the little birds hatched from them sing water songs, for they hear them all their lives, and even before they are born.

I have often observed the young just out of the nest making their odd gestures, and seeming in every way as much at home as their experienced parents, like young bees on their first excursions to the flower fields. No amount of familiarity with people and their ways seems to change them in the least. To all appearance their behavior is just the

same on seeing a man for the first time, as when
they have seen him frequently.

On the lower reaches of the rivers where mills

THE OUZEL AT HOME.

are built, they sing
on through the din
of the machinery,
and all the noisy
confusion of dogs,
cattle, and work-
men.   On one
occasion, while a
wood-chopper was at work on the river-bank, I observed one cheerily singing within reach of the flying
chips. Nor does any kind of unwonted disturbance
put him in bad humor, or frighten him out of calm
self-possession.   In passing through a narrow

gorge, I once drove one ahead of me from rapid to
rapid, disturbing him four times in quick succession where he could not very well fly past me on
account of the narrowness of the channel. Most
birds under similar circumstances fancy themselves
pursued, and become suspiciously uneasy; but, instead of growing nervous about it, he made his
usual dippings, and sang one of his most tranquil
strains. When observed within a few yards their
eyes are seen to express remarkable gentleness and
intelligence; but they seldom allow so near a view
unless one wears clothing of about the same color
as the rocks and trees, and knows how to sit still.
On one occasion, while rambling along the shore of
a mountain lake, where the birds, at least those
born that season, had never seen a man, I sat down
to rest on a large stone close to the water's edge,
upon which it seemed the ouzels and sandpipers
were in the habit of alighting when they came to
feed on that part of the shore, and some of the
other birds also, when they came down to wash or
drink. In a few minutes, along came a whirring
Ouzel and alighted on the stone beside me, within
reach of my hand. Then suddenly observing me,
he stooped nervously as if about to fly on the in-
stant, but as I remained as motionless as the stone,
he gained confidence, and looked me steadily in the
face for about a minute, then flew quietly to the
outlet and began to sing. Next came a sandpiper
and gazed at me with much the same guileless expression of eye as the Ouzel. Lastly, down with a
swoop came a Steller's jay out of a fir-tree, probably with the intention of moistening his noisy

throat. But instead of sitting confidingly as my other visitors had done, he rushed off at once, nearly tumbling heels over head into the lake in his suspicious confusion, and with loud screams roused the neighborhood.

Love for song-birds, with their sweet human voices, appears to be more common and unfailing than love for flowers. Every one loves flowers to some extent, at least in life's fresh morning, attracted by them as instinctively as humming-birds and bees. Even the young Digger Indians have sufficient love for the brightest of those found growing on the mountains to gather them and braid them as decorations for the hair. And I was glad to discover, through the few Indians that could be induced to talk on the subject, that they have names for the wild rose and the lily, and other conspicuous flowers, whether available as food or otherwise. Most men, however, whether savage or civilized, become apathetic toward all plants that have no other apparent use than the use of beauty. But fortunately one's first instinctive love of song-birds is never wholly obliterated, no matter what the influences upon our lives may be. I have often been delighted to see a pure, spiritual glow come into the countenances of hard business-men and old miners, when a song-bird chanced to alight near them. Nevertheless, the little mouthful of meat that swells out the breasts of some song-birds is too often the cause of their death. Larks and robins in particular are brought to market in hundreds. But fortunately the Ouzel has no enemy so eager to eat his little body as to follow him into the moun-

tain solitudes. I never knew him to be chased
even by hawks.

An acquaintance of mine, a sort of foot-hill
mountaineer, had a pet cat, a great, dozy, overgrown creature, about as broad-shouldered as a lynx.
During the winter, while the snow lay deep, the
mountaineer sat in his lonely cabin among the
pines smoking his pipe and wearing the dull time
away. Tom was his sole companion, sharing his
bed, and sitting beside him on a stool with much
the same drowsy expression of eye as his master.
The good-natured bachelor was content with his
hard fare of soda-bread and bacon, but Tom, the
only creature in the world acknowledging dependence on him, must needs be provided with fresh
meat. Accordingly he bestirred himself to contrive
squirrel-traps, and waded the snowy woods with
his gun, making sad havoc among the few winter
birds, sparing neither robin, sparrow, nor tiny nut-
hatch, and the pleasure of seeing Tom eat and
grow fat was his great reward.

One cold afternoon, while hunting along the
river-bank, he noticed a plain-feathered little bird
skipping about in the shallows, and immediately
raised his gun. But just then the confiding songster began to sing, and after listening to his summery melody the charmed hunter turned away,
saying, " Bless your little heart, I can't shoot you,
not even for Tom."

Even so far north as icy Alaska, I have found
my glad singer. When I was exploring the glaciers between Mount Fairweather and the Stikeen
River, one cold day in November, after trying

YOSEMITE BIRDS, SNOW-BOUND AT THE FOOT OF INDIAN CAÑON.

in vain to force a way through the innumerable
icebergs of Sum Dum Bay to the great glaciers at
the head of it, I was weary and baffled and sat
resting in my canoe convinced at last that I would
have to leave this part of my work for another
year.   Then I began to plan my escape to open
water before the young ice which was beginning to
form should shut me in.   While I thus lingered
drifting with the bergs, in the midst of these
gloomy forebodings and all the terrible glacial desolation and grandeur, I suddenly heard the well-
known whir of an Ouzel's wings, and, looking up,
saw my little comforter coming straight across the
ice from the shore.   In a second or two he was
with me, flying three times round my head with a
happy salute, as if saying, "Cheer up, old friend;
you see I 'm here, and all 's well."   Then he flew
back to the shore, alighted on the topmost jag of a
stranded iceberg, and began to nod and bow as
though he were on one of his favorite boulders in
the midst of a sunny Sierra cascade.

The species is distributed all along the mountain-
ranges of the Pacific Coast from Alaska to Mexico,
and east to the Rocky Mountains.   Nevertheless,
it is as yet comparatively little known.   Audubon
and Wilson did not meet it.   Swainson was, I believe, the first naturalist to describe a specimen
from Mexico.   Specimens were shortly afterward
procured by Drummond near the sources of the
Athabasca River, between the fifty-fourth and
fifty-sixth parallels; and it has been collected by
nearly all of the numerous exploring expeditions
undertaken of late through our Western States and

Territories; for it never fails to engage the attention of naturalists in a very particular manner.

Such, then, is our little cinclus, beloved of every one who is so fortunate as to know him. Tracing on strong wing every curve of the most precipitous torrents from one extremity of the Sierra to the other; not fearing to follow them through their darkest gorges and coldest snow-tunnels; acquainted with every waterfall, echoing their divine music; and throughout the whole of their beautiful lives interpreting all that we in our unbelief call terrible in the utterances of torrents and storms, as only varied expressions of God's eternal love.

CHAPTER XIV

THE wild sheep ranks highest among the animal mountaineers of the Sierra. Possessed of keen sight and scent, and strong limbs, he dwells secure amid the loftiest summits, leaping unscathed from crag to crag, up and down the fronts of giddy precipices, crossing foaming torrents and slopes of frozen snow, exposed to the wildest storms, yet maintaining a brave, warm life, and developing from generation to generation in perfect strength and beauty.

Nearly all the lofty mountain-chains of the globe are inhabited by wild sheep, most of which, on account of the remote and all but inaccessible regions where they dwell, are imperfectly known as yet. They are classified by different naturalists under from five to ten distinct species or varieties, the best known being the burrhel of the Himalaya (*Ovis burrhel*, Blyth); the argali, the large wild sheep of central and northeastern Asia (*O. ammon*, Linn., or *Caprovis argali*); the Corsican mouflon (*O. musimon*, Pal.); the aoudad of the mountains of northern Africa (*Ammotragus tragelaphus*); and the Rocky Mountain bighorn (*O. montana*, Cuv.).

300

To this last-named species belongs the wild sheep of the Sierra. Its range, according to the late Professor Baird of the Smithsonian Institution, extends "from the region of the upper Missouri and Yellowstone to the Rocky Mountains and the high grounds adjacent to them on the eastern slope, and as far south as the Rio Grande. Westward it extends to the coast ranges of Washington, Oregon, and California, and follows the highlands some distance into Mexico." [1] Throughout the vast region bounded on the east by the Wahsatch Mountains and on the west by the Sierra there are more than a hundred subordinate ranges and mountain groups, trending north and south, range beyond range, with summits rising from eight to twelve thousand feet above the level of the sea, probably all of which, according to my own observations, is, or has been, inhabited by this species.

Compared with the argali, which, considering its size and the vast extent of its range, is probably the most important of all the wild sheep, our species is about the same size, but the horns are less twisted and less divergent. The more important characteristics are, however, essentially the same, some of the best naturalists maintaining that the two are only varied forms of one species. In accordance with this view, Cuvier conjectures that since central Asia seems to be the region where the sheep first appeared, and from which it has been distributed, the argali may have been distributed over this continent from Asia by crossing Bering Strait on ice. This conjecture is not so ill

1 Pacific Railroad Survey, Vol. VIII, page 678.

founded as at first sight would appear; for the Strait is only about fifty miles wide, is interrupted by three islands, and is jammed with ice nearly every winter. Furthermore the argali is abundant on the mountains adjacent to the Strait at East Cape, where it is well known to the Tschuckchi hunters and where I have seen many of their horns.

On account of the extreme variability of the sheep under culture, it is generally supposed that the innumerable domestic breeds have all been derived from the few wild species; but the whole question is involved in obscurity. According to Darwin, sheep have been domesticated from a very ancient period, the remains of a small breed, differing from any now known, having been found in the famous Swiss lake-dwellings.

Compared with the best-known domestic breeds, we find that our wild species is much larger, and, instead of an all-wool garment, wears a thick over-coat of hair like that of the deer, and an under-covering of fine wool. The hair, though rather coarse, is comfortably soft and spongy, and lies smooth, as if carefully tended with comb and brush. The predominant color during most of the year is brownish-gray, varying to bluish-gray in the autumn; the belly and a large, conspicuous patch on the buttocks are white; and the tail, which is very short, like that of a deer, is black, with a yellowish border. The wool is white, and grows in beautiful spirals down out of sight among the shining hair, like delicate climbing vines among stalks of corn.

The horns of the male are of immense size, measuring in their greater diameter from five to six and a half inches, and from two and a half to three feet in length around the curve. They are yellow-ish-white in color, and ridged transversely, like those of the domestic ram. Their cross-section near the base is somewhat triangular in outline, and flattened toward the tip. Rising boldly from the top of the head, they curve gently backward and outward, then forward and outward, until about three fourths of a circle is described, and until the flattened, blunt tips are about two feet or two and a half feet apart. Those of the female are flattened throughout their entire length, are less curved than those of the male, and much smaller, measuring less than a foot along the curve.

A ram and ewe that I obtained near the Modoc lava-beds, to the northeast of Mount Shasta, measured as follows:

	Ram.		Ewe.	
	ft.	in.	ft.	in.
Height at shoulders	3	6	3	0
Girth around shoulders	3	11	3	$3\frac{3}{4}$
Length from nose to root of tail	5	$10\frac{1}{4}$	4	$3\frac{1}{2}$
Length of ears	0	$4\frac{3}{4}$	0	5
Length of tail	0	$4\frac{1}{2}$	0	$4\frac{1}{2}$
Length of horns around curve	2	9	0	$11\frac{1}{2}$
Distance across from tip to tip of horns	2	$5\frac{1}{2}$		
Circumference of horns at base	1	4	0	6

The measurements of a male obtained in the Rocky Mountains by Audubon vary but little as compared with the above. The weight of his speci-men was 344 pounds,[1] which is, perhaps, about an

[1] Audubon and Bachman's "Quadrupeds of North America."

average for full-grown males. The females are about a third lighter.

Besides these differences in size, color, hair, etc., as noted above, we may observe that the domestic sheep, in a general way, is expressionless, like a dull bundle of something only half alive, while the wild is as elegant and graceful as a deer, every movement manifesting admirable strength and character. The tame is timid; the wild is bold. The tame is always more or less ruffled and dirty; while the wild is as smooth and clean as the flowers of his mountain pastures.

The earliest mention that I have been able to find of the wild sheep in America is by Father Picolo, a Catholic missionary at Monterey, in the year 1797, who, after describing it, oddly enough, as "a kind of deer with a sheep-like head, and about as large as a calf one or two years old," naturally hurries on to remark: "I have eaten of these beasts; their flesh is very tender and delicious." Mackenzie, in his northern travels, heard the species spoken of by the Indians as "white buffaloes." And Lewis and Clark tell us that, in a time of great scarcity on the head waters of the Missouri, they saw plenty of wild sheep, but they were "too shy to be shot."

A few of the more energetic of the Pah Ute Indians hunt the wild sheep every season among the more accessible sections of the High Sierra, in the neighborhood of passes, where, from having been pursued, they have become extremely wary; but in the rugged wilderness of peaks and cañons, where the foaming tributaries of the San Joaquin and King's rivers take their rise, they fear no hunter

save the wolf, and are more guileless and approach-
able than their tame kindred.

While engaged in the work of exploring high regions where they delight to roam I have been greatly
interested in studying their habits. In the months
of November and December, and probably during
a considerable portion of midwinter, they all flock
together, male and female, old and young. I once
found a complete band of this kind numbering upward of fifty, which, on being alarmed, went bounding away across a jagged lava-bed at admirable
speed, led by a majestic old ram, with the lambs
safe in the middle of the flock.

In spring and summer, the full-grown rams form
separate bands of from three to twenty, and are
usually found feeding along the edges of glacier
meadows, or resting among the castle-like crags of
the high summits; and whether quietly feeding, or
scaling the wild cliffs, their noble forms and the
power and beauty of their movements never fail
to strike the beholder with lively admiration.

Their resting-places seem to be chosen with reference to sunshine and a wide outlook, and most
of all to safety. Their feeding-grounds are among
the most beautiful of the wild gardens, bright with
daisies and gentians and mats of purple bryanthus, lying hidden away on rocky headlands and
cañon sides, where sunshine is abundant, or down
in the shady glacier valleys, along the banks of
the streams and lakes, where the plushy sod is
greenest. Here they feast all summer, the happy
wanderers, perhaps relishing the beauty as well as
the taste of the lovely flora on which they feed.

20

When the winter storms set in, loading their
highland pastures with snow, then, like the birds,

SNOW-BOUND ON MOUNT SHASTA.

they gather and go to lower climates, usually descending the eastern flank of the range to the rough,
volcanic table-lands and treeless ranges of the

Great Basin adjacent to the Sierra. They never make haste, however, and seem to have no dread of storms, many of the strongest only going down leisurely to bare, wind-swept ridges, to feed on bushes and dry bunch-grass, and then returning up into the snow. Once I was snow-bound on Mount Shasta for three days, a little below the timber line. It was a dark and stormy time, well calculated to test the skill and endurance of mountaineers. The snow-laden gale drove on night and day in hissing, blinding floods, and when at length it began to abate, I found that a small band of wild sheep had weathered the storm in the lee of a clump of Dwarf Pines a few yards above my storm-nest, where the snow was eight or ten feet deep. I was warm back of a rock, with blankets, bread, and fire. My brave companions lay in the snow, without food, and with only the partial shelter of the short trees, yet they made no sign of suffering or faint-heartedness.

In the months of May and June, the wild sheep bring forth their young in solitary and almost inaccessible crags, far above the nesting-rocks of the eagle. I have frequently come upon the beds of the ewes and lambs at an elevation of from 12,000 to 13,000 feet above sea-level. These beds are simply oval-shaped hollows, pawed out among loose, disintegrating rock-chips and sand, upon some sunny spot commanding a good outlook, and partially sheltered from the winds that sweep those lofty peaks almost without intermission. Such is the cradle of the little mountaineer, aloft in the very sky; rocked in storms, curtained in clouds, sleeping in thin, icy air; but, wrapped in his hairy coat, and nourished by a strong, warm mother, defended from the talons of the eagle and the teeth of the sly coyote, the bonny lamb grows apace. He soon learns to nibble the tufted rock-grasses and leaves of the white spiræa; his horns begin to shoot, and before summer is done he is strong and agile, and goes forth with the flock, watched by the same divine love that tends the more helpless human lamb in its cradle by the fireside.

Nothing is more commonly remarked by noisy, dusty trail-travelers in the Sierra than the want of animal life—no song-birds, no deer, no squirrels, no game of any kind, they say. But if such could only go away quietly into the wilderness, sauntering afoot and alone with natural deliberation, they would soon learn that these mountain mansions are not without inhabitants, many of whom, confiding and gentle, would not try to shun their acquaintance.

In the fall of 1873 I was tracing the South Fork of the San Joaquin up its wild cañon to its farthest glacier fountains. It was the season of alpine Indian summer. The sun beamed lovingly; the squirrels were nutting in the pine-trees, butterflies hovered about the last of the goldenrods, the willow and maple thickets were yellow, the meadows brown, and the whole sunny, mellow landscape glowed like a countenance in the deepest and sweetest repose. On my way over the glacier-polished rocks along the river, I came to an expanded portion of the cañon, about two miles long and half a mile wide, which formed a level park inclosed with picturesque granite walls like those of Yosemite Valley. Down through the middle of it

poured the beautiful river shining and spangling
in the golden light, yellow groves on its banks, and
strips of brown meadow; while the whole park was
astir with wild life, some of which even the noisiest

HEAD OF THE MERINO RAM (DOMESTIC).

and least observing of travelers must have seen had
they been with me. Deer, with their supple, well-
grown fawns, bounded from thicket to thicket as I
advanced; grouse kept rising from the brown grass
with a great whirring of wings, and, alighting on

the lower branches of the pines and poplars, allowed a near approach, as if curious to see me.
Farther on, a broad-shouldered wildcat showed
himself, coming out of a grove, and crossing the
river on a flood-jamb of logs, halting for a moment
to look back. The bird-like tamias frisked about
my feet everywhere among the pine-needles and
seedy grass-tufts; cranes waded the shallows of the
river-bends, the kingfisher rattled from perch to
perch, and the blessed ouzel sang amid the spray
of every cascade. Where may lonely wanderer
find a more interesting family of mountain-dwell-
ers, earth-born companions and fellow-mortals?
It was afternoon when I joined them, and the glorious landscape began to fade in the gloaming before I awoke from their enchantment. Then I
sought a camp-ground on the river-bank, made a
cupful of tea, and lay down to sleep on a smooth
place among the yellow leaves of an aspen grove.
Next day I discovered yet grander landscapes and
grander life. Following the river over huge, swelling rock-bosses through a majestic cañon, and past
innumerable cascades, the scenery in general became gradually wilder and more alpine. The Sugar Pine and Silver Firs gave place to the hardier
Cedar and Hemlock Spruce. The cañon walls became more rugged and bare, and gentians and arctic daisies became more abundant in the gardens
and strips of meadow along the streams. Toward
the middle of the afternoon I came to another valley, strikingly wild and original in all its features,
and perhaps never before touched by human foot.
As regards area of level bottom-land, it is one of

the very smallest of the Yosemite type, but its walls
are sublime, rising to a height of from 2000 to 4000
feet above the river. At the head of the valley
the main cañon forks, as is found to be the case in
all yosemites. The formation of this one is due

HEAD OF ROCKY MOUNTAIN WILD SHEEP.

chiefly to the action of two great glaciers, whose
fountains lay to the eastward, on the flanks of
Mounts Humphrey and Emerson and a cluster of
nameless peaks farther south.

The gray, boulder-chafed river was singing loudly through the valley, but above its massy roar I heard the booming of a waterfall, which drew me eagerly on; and just as I emerged from the tangled groves and brier-thickets at the head of the valley, the main fork of the river came in sight, falling fresh from its glacier fountains in a snowy cascade, between granite walls 2000 feet high. The steep incline down which the glad waters thundered seemed to bar all farther progress. It was not long, however, before I discovered a crooked seam in the rock, by which I was enabled to climb to the edge of a terrace that crosses the cañon, and divides the cataract nearly .in the middle. Here I sat down to take breath and make some entries in my note-book, taking advantage, at the same time, of my elevated position above the trees to gaze back over the valley into the heart of the noble landscape, little knowing the while what neighbors were near.

After spending a few minutes in this way, I chanced to look across the fall, and there stood three sheep quietly observing me. Never did the sudden appearance of a mountain, or fall, or human friend more forcibly seize and rivet my attention. Anxiety to observe accurately held me perfectly still. Eagerly I marked the flowing undulations of their firm, braided muscles, their strong legs, ears, eyes, heads, their graceful rounded necks, the color of their hair, and the bold, up-sweeping curves of their noble horns. When they moved I watched every gesture, while they, in no wise disconcerted either by my attention or by the tumultuous roar of the water, advanced deliberately alongside the rapids, between the two divisions of the cataract, turning now and then to look at me. Presently they came to a steep, ice-burnished acclivity, which they ascended by a succession of quick, short, stiff-legged leaps, reaching the top without a struggle. This was the most startling feat of mountaineering I had ever witnessed, and, considering only the mechanics of the thing, my astonishment could hardly have been greater had they displayed wings and taken to flight. " Sure-footed" mules on such ground would have fallen and rolled like loosened boulders. Many a time, where the slopes are far lower, I have been compelled to take off my shoes and stockings, tie them to my belt, and creep barefooted, with the utmost caution. No wonder then, that I watched the progress of these animal mountaineers with keen sympathy, and exulted in the boundless sufficiency of wild nature displayed in their invention, construction, and keeping. A few minutes later I caught sight of a dozen more in one band, near the foot of the upper fall. They were standing on the same side of the river with me, only twenty-five or thirty yards away, looking as unworn and perfect as if created on the spot. It appeared by their tracks, which I had seen in the Little Yosemite, and by their present position, that when I came up the cañon they were all feeding together down in the valley, and in their haste to reach high ground, where they could look about them to ascertain the nature of the strange disturbance, they were divided, three ascending on one side the river, the rest on the other.

CROSSING A CAÑON STREAM.

The main band, headed by an experienced chief,
now began to cross the wild rapids between the
two divisions of the cascade. This was another

exciting feat; for, among all the varied experiences of mountaineers, the crossing of boisterous, rock-dashed torrents is found to be one of the most trying to the nerves. Yet these fine fellows walked fearlessly to the brink, and jumped from boulder to boulder, holding themselves in easy poise above the whirling, confusing current, as if they were doing nothing extraordinary.

In the immediate foreground of this rare picture there was a fold of ice-burnished granite, traversed by a few bold lines in which rock-ferns and tufts of bryanthus were growing, the gray cañon walls on the sides, nobly sculptured and adorned with brown cedars and pines; lofty peaks in the distance, and in the middle ground the snowy fall, the voice and soul of the landscape; fringing bushes beating time to its thunder-tones, the brave sheep in front of it, their gray forms slightly obscured in the spray, yet standing out in good, heavy relief against the close white water, with their huge horns rising like the upturned roots of dead pine-trees, while the evening sunbeams streaming up the cañon colored all the picture a rosy purple and made it glorious. After crossing the river, the dauntless climbers, led by their chief, at once began to scale the cañon wall, turning now right, now left, in long, single file, keeping well apart out of one another's way, and leaping in regular succession from crag to crag, now ascending slippery dome-curves, now walking leisurely along the edges of precipices, stopping at times to gaze down at me from some flat-topped rock, with heads held aslant, as if curious to learn what I thought about it, or whether I was likely to follow them. After reaching the top of the wall, which, at this place, is somewhere between 1500 and 2000 feet high, they were still visible against the sky as they lingered, looking down in groups of twos or threes.

Throughout the entire ascent they did not make a single awkward step, or an unsuccessful effort of any kind. I have frequently seen tame sheep in mountains jump upon a sloping rock-surface, hold on tremulously a few seconds, and fall back baffled and irresolute. But in the most trying situations, where the slightest want or inaccuracy would have been fatal, these always seemed to move in comfortable reliance on their strength and skill, the limits of which they never appeared to know. Moreover, each one of the flock, while following the guidance of the most experienced, yet climbed with intelligent independence as a perfect individual, capable of separate existence whenever it should wish or be compelled to withdraw from the little clan. The domestic sheep, on the contrary, is only a fraction of an animal, a whole flock being required to form an individual, just as numerous flowerets are required to make one complete sunflower.

Those shepherds who, in summer, drive their flocks to the mountain pastures, and, while watching them night and day, have seen them frightened by bears and storms, and scattered like wind-driven chaff, will, in some measure, be able to appreciate the self-reliance and strength and noble individuality of Nature's sheep.

Like the Alp-climbing ibex of Europe, our mountaineer is said to plunge headlong down the faces

of sheer precipices, and alight on his big horns.
I know only two hunters who claim to have actually witnessed this feat; I never was so fortunate.
They describe the act as a diving head-foremost.
The horns are so large at the base that they cover
the upper portion of the head down nearly to a
level with the eyes, and the skull is exceedingly
strong. I struck an old, bleached specimen on
Mount Ritter a dozen blows with my ice-ax without breaking it. Such skulls would not fracture
very readily by the wildest rock-diving, but other
bones could hardly be expected to hold together in
such a performance; and the mechanical difficulties in the way of controlling their movements,
after striking upon an irregular surface, are, in
themselves, sufficient to show this boulder-like
method of progression to be impossible, even in
the absence of all other evidence on the subject;
moreover, the ewes follow wherever the rams may
lead, although their horns are mere spikes. I have
found many pairs of the horns of the old rams considerably battered, doubtless a result of fighting.
I was particularly interested in the question, after
witnessing the performances of this San Joaquin
band upon the glaciated rocks at the foot of the
falls; and as soon as I procured specimens and
examined their feet, all the mystery disappeared.
The secret, considered in connection with exceptionally strong muscles, is simply this: the wide
posterior portion of the bottom of the foot, instead
of wearing down and becoming flat and hard, like
the feet of tame sheep and horses, bulges out in a
soft, rubber-like pad or cushion, which not only

grips and holds well on smooth rocks, but fits into small cavities, and down upon or against slight protuberances. Even the hardest portions of the edge of the hoof are comparatively soft and elastic; furthermore, the toes admit of an extraordinary amount of both lateral and vertical movement, allowing the foot to accommodate itself still more perfectly to the irregularities of rock surfaces, while at the same time increasing the gripping power.

At the base of Sheep Rock, one of the winter strongholds of the Shasta flocks, there lives a stock-raiser who has had the advantage of observing the movements of wild sheep every winter; and, in the course of a conversation with him on the subject of their diving habits, he pointed to the front of a lava headland about 150 feet high, which is only eight or ten degrees out of the perpendicular. "There," said he, "I followed a band of them fellows to the back of that rock yonder, and expected to capture them all, for I thought I had a dead thing on them. I got behind them on a narrow bench that runs along the face of the wall near the top and comes to an end where they could n't get away without falling and being killed; but they jumped off, and landed all right, as if that were the regular thing with them."

"What!" said I, "jumped 150 feet perpendicular! Did you see them do it?"

"No," he replied, "I did n't see them going down, for I was behind them; but I saw them go off over the brink, and then I went below and found their tracks where they struck on the loose rubbish at the bottom. They just *sailed right off*, and landed on their feet right side up. That is the kind of animal *they* is—beats anything else that goes on four legs."

On another occasion, a flock that was pursued by hunters retreated to another portion of this same cliff where it is still higher, and, on being followed, they were seen jumping down in perfect order, one behind another, by two men who happened to be chopping where they had a fair view of them and could watch their progress from top to bottom of the precipice. Both ewes and rams made the frightful descent without evincing any extraordinary concern, hugging the rock closely, and controlling the velocity of their half falling,

WILD SHEEP JUMPING OVER A PRECIPICE.

half leaping movements by striking at short intervals and holding back with their cushioned, rubber feet upon small ledges and roughened inclines until near the bottom, when they "sailed off" into the free air and alighted on their feet, but with their bodies so nearly in a vertical position that they appeared to be diving.

It appears, therefore, that the methods of this wild mountaineering become clearly comprehensible as soon as we make ourselves acquainted with the rocks, and the kind of feet and muscles brought to bear upon them.

The Modoc and Pah Ute Indians are, or rather have been, the most successful hunters of the wild sheep in the regions that have come under my own observation. I have seen large numbers of heads and horns in the caves of Mount Shasta and the Modoc lava-beds, where the Indians had been feasting in stormy weather; also in the cañons of the Sierra opposite Owen's Valley; while the heavy obsidian arrow-heads found on some of the highest peaks show that this warfare has long been going on.

In the more accessible ranges that stretch across the desert regions of western Utah and Nevada, considerable numbers of Indians used to hunt in company like packs of wolves, and being perfectly acquainted with the topography of their hunting-grounds, and with the habits and instincts of the game, they were pretty successful. On the tops of nearly every one of the Nevada mountains that I have visited, I found small, nest-like inclosures built of stones, in which, as I afterward learned, one or more Indians would lie in wait while their companions scoured the ridges below, knowing that the alarmed sheep would surely run to the summit, and when they could be made to approach with the wind they were shot at short range.

INDIANS HUNTING WILD SHEEP.

Still larger bands of Indians used to make extensive hunts upon some dominant mountain much frequented by the sheep, such as Mount Grant on the Wassuck Range to the west of Walker Lake. On some particular spot, favorably situated with reference to the well-known trails of the sheep, they built a high-walled corral, with long guiding wings diverging from the gateway; and into this inclosure they sometimes succeeded in driving the noble game. Great numbers of Indians were of course required, more, indeed, than they could usually muster, counting in squaws, children, and all; they were compelled, therefore, to build rows of dummy hunters out of stones, along the ridge-tops which they wished to prevent the sheep from crossing. And, without discrediting the sagacity of the game, these dummies were found effective; for, with a few live Indians moving about excitedly among them, they could hardly be distinguished at a little distance from men, by any one not in the secret. The whole ridge-top then seemed to be alive with hunters.

The only animal that may fairly be regarded as a companion or rival of the sheep is the so-called Rocky Mountain goat (*Aplocerus montana*, Rich.), which, as its name indicates, is more antelope than goat. He, too, is a brave and hardy climber, fearlessly crossing the wildest summits, and braving the severest storms, but he is shaggy, short-legged, and much less dignified in demeanor than the sheep. His jet-black horns are only about five or six inches in length, and the long, white hair with which he is covered obscures the expression of his limbs. I have never yet seen a single specimen in the Sierra, though possibly a few flocks may have lived on Mount Shasta a comparatively short time ago.

The ranges of these two mountaineers are pretty

distinct, and they see but little of each other; the
sheep being restricted mostly to the dry, inland
mountains; the goat or chamois to the wet, snowy
glacier-laden mountains of the northwest coast
of the continent in Oregon, Washington, British
Columbia, and Alaska. Probably more than 200
dwell on the icy, volcanic cone of Mount Rainier;
and while I was exploring the glaciers of Alaska I
saw flocks of these admirable mountaineers nearly
every day, and often followed their trails through
the mazes of bewildering crevasses, in which they
are excellent guides.

Three species of deer are found in California,—
the black-tailed, white-tailed, and mule deer. The
first mentioned (*Cervus Columbianus*) is by far
the most abundant, and occasionally meets the
sheep during the summer on high glacier meadows,
and along the edge of the timber line; but being
a forest animal, seeking shelter and rearing its
young in dense thickets, it seldom visits the wild
sheep in its higher homes. The antelope, though
not a mountaineer, is occasionally met in winter
by the sheep while feeding along the edges of the
sage-plains and bare volcanic hills to the east of
the Sierra. So also is the mule deer, which is almost restricted in its range to this eastern region.
The white-tailed species belongs to the coast ranges.

Perhaps no wild animal in the world is without
enemies, but highlanders, as a class, have fewer
than lowlanders. The wily panther, slipping and
crouching among long grass and bushes, pounces
upon the antelope and deer, but seldom crosses the
bald, craggy thresholds of the sheep. Neither can

the bears be regarded as enemies; for, though they seek to vary their every-day diet of nuts and berries by an occasional meal of mutton, they prefer to hunt tame and helpless flocks. Eagles and coyotes, no doubt, capture an unprotected lamb at times, or some unfortunate beset in deep, soft snow, but these cases are little more than accidents. So, also, a few perish in long-continued snow-storms, though, in all my mountaineering, I have not found more than five or six that seemed to have met their fate in this way. A little band of three were discovered snow-bound in Bloody Cañon a few years ago, and were killed with an ax by mountaineers, who chanced to be crossing the range in winter.

Man is the most dangerous enemy of all, but even from him our brave mountain-dweller has little to fear in the remote solitudes of the High Sierra. The golden plains of the Sacramento and San Joaquin were lately thronged with bands of elk and antelope, but, being fertile and accessible, they were required for human pastures. So, also, are many of the feeding-grounds of the deer — hill, valley, forest, and meadow — but it will be long before man will care to take the highland castles of the sheep. And when we consider here how rapidly entire species of noble animals, such as the elk, moose, and buffalo, are being pushed to the very verge of extinction, all lovers of wildness will rejoice with me in the rocky security of *Ovis montana*, the bravest of all the Sierra mountaineers.

CHAPTER XV

IN THE SIERRA FOOT-HILLS

MURPHY'S CAMP is a curious old mining-town in Calaveras County, at an elevation of 2400 feet above the sea, situated like a nest in the center of a rough, gravelly region, rich in gold. Granites, slates, lavas, limestone, iron ores, quartz veins, auriferous gravels, remnants of dead fire-rivers and dead water-rivers are developed here side by side within a radius of a few miles, and placed invitingly open before the student like a book, while the people and the region beyond the camp furnish mines of study of never-failing interest and variety.

When I discovered this curious place, I was tracing the channels of the ancient pre-glacial rivers, instructive sections of which have been laid bare here and in the adjacent regions by the miners. Rivers, according to the poets, "go on forever"; but those of the Sierra are young as yet and have scarcely learned the way down to the sea; while at least one generation of them have died and vanished together with most of the basins they drained. All that remains of them to tell their history is a series of interrupted fragments of channels, mostly choked with gravel, and buried beneath broad,

thick sheets of lava. These are known as the
"Dead Rivers of California," and the gravel deposited in them is comprehensively called the
"Blue Lead." In some places the channels of the
present rivers trend in the same direction, or
nearly so, as those of the ancient rivers; but, in
general, there is little correspondence between
them, the entire drainage having been changed, or,
rather, made new. Many of the hills of the ancient landscapes have become hollows, and the
old hollows have become hills. Therefore the
fragmentary channels, with their loads of auriferous gravel, occur in all kinds of unthought-of
places, trending obliquely, or even at right angles
to the present drainage, across the tops of lofty
ridges or far beneath them, presenting impressive
illustrations of the magnitude of the changes accomplished since those ancient streams were annihilated. The last volcanic period preceding the
regeneration of the Sierra landscapes seems to
have come on over all the range almost simultaneously, like the glacial period, notwithstanding
lavas of different age occur together in many
places, indicating numerous periods of activity in
the Sierra fire-fountains. The most important of
the ancient river-channels in this region is a section that extends from the south side of the town
beneath Coyote Creek and the ridge beyond it
to the Cañon of the Stanislaus; but on account of
its depth below the general surface of the present
valleys the rich gold gravels it is known to contain
cannot be easily worked on a large scale. Their
extraordinary richness may be inferred from the

fact that many claims were profitably worked in them by sinking shafts to a depth of 200 feet or more, and hoisting the dirt by a windlass. Should the dip of this ancient channel be such as to make the Stanislaus Cañon available as a dump, then the grand deposit might be worked by the hydraulic method, and although a long, expensive tunnel would be required, the scheme might still prove profitable, for there is "millions in it."

The importance of these ancient gravels as gold fountains is well known to miners. Even the superficial placers of the present streams have derived much of their gold from them. According to all accounts, the Murphy placers have been very rich —"terrific rich," as they say here. The hills have been cut and scalped, and every gorge and gulch and valley torn to pieces and disemboweled, expressing a fierce and desperate energy hard to understand. Still, any kind of effort-making is better than inaction, and there is something sublime in seeing men working in dead earnest at anything, pursuing an object with glacier-like energy and persistence. Many a brave fellow has recorded a most eventful chapter of life on these Calaveras rocks. But most of the pioneer miners are sleeping now, their wild day done, while the few survivors linger languidly in the washed-out gulches or sleepy village like harried bees around the ruins of their hive. "We have no industry left *now*," they told me, "and no men; everybody and everything hereabouts has gone to decay. We are only bummers — out of the game, a thin scatterin' of poor, dilapidated cusses, compared with what we used to be in the grand old gold-days. We were giants then, and you can look around here and see our tracks." But although these lingering pioneers are perhaps more exhausted than the mines, and about as dead as the dead rivers, they are yet a rare and interesting set of men, with much gold mixed with the rough, rocky gravel of their characters; and they manifest a breeding and intelligence little looked for in such surroundings as theirs. As the heavy, long-continued grinding of the glaciers brought out the features of the Sierra, so the intense experiences of the gold period have brought out the features of these old miners, forming a richness and variety of character little known as yet. The sketches of Bret Harte, Hayes, and Miller have not exhausted this field by any means. It is interesting to note the extremes possible in one and the same character: harshness and gentleness, manliness and childishness, apathy and fierce endeavor. Men who, twenty years ago, would not cease their shoveling to save their lives, now play in the streets with children. Their long, Micawber-like waiting after the exhaustion of the placers has brought on an exaggerated form of dotage. I heard a group of brawny pioneers in the street eagerly discussing the quantity of tail required for a boy's kite; and one graybeard undertook the sport of flying it, volunteering the information that he was a boy, "always was a boy, and d—n a man who was not a boy inside, however ancient outside!" Mines, morals, politics, the immortality of the soul, etc., were discussed beneath shade-trees and in saloons, the time for each being gov-

erned apparently by the temperature. Contact with
Nature, and the habits of observation acquired in
gold-seeking, had made them all, to some extent,
collectors, and, like wood-rats, they had gathered
all kinds of odd specimens into their cabins, and
now required me to examine them. They were
themselvès the oddest and most interesting specimens. One of them offered to show me around
the old diggings, giving me fair warning before
setting out that I might not like him, "because,"
said he, "people say I 'm eccentric. I notice
everything, and gather beetles and snakes and
anything that 's queer; and so some don't like me,
and call me eccentric. I 'm always trying to find
out things. Now, there 's a weed; the Indians eat
it for greens. What do you call those long-bodied
flies with big heads?" "Dragon-flies," I suggested.
"Well, their jaws work sidewise, instead of up and
down, and grasshoppers' jaws work the same way,
and therefore I think they are the same species.
I always notice everything like that, and just because I do, they say I 'm eccentric," etc.

Anxious that I should miss none of the wonders
of their old gold-field, the good people had much to
say about the marvelous beauty of Cave City Cave,
and advised me to explore it. This I was very
glad to do, and finding a guide who knew the way
to the mouth of it, I set out from Murphy the next
morning.

The most beautiful and extensive of the mountain caves of California occur in a belt of metamorphic limestone that is pretty generally developed
along the western flank of the Sierra from the Mc-

Cloud River on the north to the Kaweah on the south, a distance of over 400 miles, at an elevation of from 2000 to 7000 feet above the sea. Besides this regular belt of caves, the California landscapes are diversified by long imposing ranks of sea-caves, rugged and variable in architecture, carved in the coast headlands and precipices by centuries of wave-dashing; and innumerable lava-caves, great and small, originating in the unequal flowing and hardening of the lava sheets in which they occur, fine illustrations of which are presented in the famous Modoc Lava Beds, and around the base of icy Shasta. In this comprehensive glance we may also notice the shallow wind-worn caves in stratified sandstones along the margins of the plains; and the cave-like recesses in the Sierra slates and granites, where bears and other mountaineers find shelter during the fall of sudden storms. In general, however, the grand massive uplift of the Sierra, as far as it has been laid bare to observation, is about as solid and caveless as a boulder.

Fresh beauty opens one's eyes wherever it is really seen, but the very abundance and completeness of the common beauty that besets our steps prevents its being absorbed and appreciated. It is a good thing, therefore, to make short excursions now and then to the bottom of the sea among dulse and coral, or up among the clouds on mountain-tops, or in balloons, or even to creep like worms into dark holes and caverns underground, not only to learn something of what is going on in those out-of-the-way places, but to see better what the sun sees on our return to common every-day beauty.

Our way from Murphy's to the cave lay across a series of picturesque, moory ridges in the chaparral region between the brown foot-hills and the forests, a flowery stretch of rolling hill-waves breaking here and there into a kind of rocky foam on the higher summits, and sinking into delightful bosky hollows embowered with vines. The day was a fine specimen of California summer, pure sunshine, unshaded most of the time by a single cloud. As the sun rose higher, the heated air began to flow in tremulous waves from every southern slope. The sea-breeze that usually comes up the foot-hills at this season, with cooling on its wings, was scarcely perceptible. The birds were assembled beneath leafy shade, or made short, languid flights in search of food, all save the majestic buzzard; with broad wings outspread he sailed the warm air unwearily from ridge to ridge, seeming to enjoy the fervid sunshine like a butterfly. Squirrels, too, whose spicy ardor no heat or cold may abate, were nutting among the pines, and the innumerable hosts of the insect kingdom were throbbing and wavering unwearied as sunbeams.

This brushy, berry-bearing region used to be a deer and bear pasture, but since the disturbances of the gold period these fine animals have almost wholly disappeared. Here, also, once roamed the mastodon and elephant, whose bones are found entombed in the river gravels and beneath thick folds of lava. Toward noon, as we were riding slowly over bank and brae, basking in the unfeverish sun-heat, we witnessed the upheaval of a new mountain-range, a Sierra of clouds abounding in land-

scapes as truly sublime and beautiful—if only we
have a mind to think so and eyes to see—as the
more ancient rocky Sierra beneath it, with its forests and waterfalls; reminding us that, as there is a
lower world of caves, so, also, there is an upper
world of clouds. Huge, bossy cumuli developed
with astonishing rapidity from mere buds, swelling
with visible motion into colossal mountains, and
piling higher, higher, in long massive ranges, peak
beyond peak, dome over dome, with many a picturesque valley and shadowy cave between; while
the dark firs and pines of the upper benches of the
Sierra were projected against their pearl bosses
with exquisite clearness of outline. These cloud
mountains vanished in the azure as quickly as they
were developed, leaving no detritus; but they were
not a whit less real or interesting on this account.
The more enduring hills over which we rode were
vanishing as surely as they, only not so fast, a difference which is great or small according to the
standpoint from which it is contemplated.

At the bottom of every dell we found little home-
steads embosomed in wild brush and vines wherever the recession of the hills left patches of arable
ground. These secluded flats are settled mostly by
Italians and Germans, who plant a few vegetables
and grape-vines at odd times, while their main
business is mining and prospecting. In spite of all
the natural beauty of these dell cabins, they can
hardly be called homes. They are only a better
kind of camp, gladly abandoned whenever the
hoped-for gold harvest has been gathered. There is
an air of profound unrest and melancholy about

the best of them. Their beauty is thrust upon them by exuberant Nature, apart from which they are only a few logs and boards rudely jointed and without either ceiling or floor, a rough fireplace with corresponding cooking utensils, a shelf-bed, and stool. The ground about them is strewn with battered prospecting-pans, picks, sluice-boxes, and quartz specimens from many a ledge, indicating the trend of their owners' hard lives.

The ride from Murphy's to the cave is scarcely two hours long, but we lingered among quartz-ledges and banks of dead river gravel until long after noon. At length emerging from a narrow-throated gorge, a small house came in sight set in a thicket of fig-trees at the base of a limestone hill. "That," said my guide, pointing to the house, "is Cave City, and the cave is in that gray hill." Arriving at the one house of this one-house city, we were boisterously welcomed by three drunken men who had come to town to hold a spree. The mistress of the house tried to keep order, and in reply to our inquiries told us that the cave guide was then in the cave with a party of ladies. "And must we wait until he returns?" we asked. No, that was unnecessary; we might take candles and go into the cave alone, provided we shouted from time to time so as to be found by the guide, and were careful not to fall over the rocks or into the dark pools. Accordingly taking a trail from the house, we were led around the base of the hill to the mouth of the cave, a small inconspicuous archway, mossy around the edges and shaped like the door of a water-ouzel's nest, with no appreciable hint or advertisement of

the grandeur of the many crystal chambers within.
Lighting our candles, which seemed to have no illu-
minating power in the thick darkness, we groped
our way onward as best we could along narrow
lanes and alleys, from chamber to chamber, around
rustic columns and heaps of fallen rocks, stopping
to rest now and then in particularly beautiful
places — fairy alcoves furnished with admirable variety of shelves and tables, and round bossy stools
covered with sparkling crystals. Some of the cor-
ridors were muddy, and in plodding along these we
seemed to be in the streets of some prairie village
in spring-time. Then we would come to handsome
marble stairways conducting right and left into
upper chambers ranged above one another three or
four stories high, floors, ceilings, and walls lavishly
decorated with innumerable crystalline forms.
After thus wandering exploringly, and alone for a
mile or so, fairly enchanted, a murmur of voices
and a gleam of light betrayed the approach of the
guide and his party, from whom, when they came
up, we received a most hearty and natural stare, as
we stood half concealed in a side recess among
stalagmites. I ventured to ask the dripping, crouching company how they had enjoyed their saunter,
anxious to learn how the strange sunless scenery of
the underworld had impressed them. "Ah, it's
nice! It's splendid!" they all replied and echoed.
"The Bridal Chamber back here is just glorious!
This morning we came down from the Calaveras
Big Tree Grove, and the trees are nothing to it."
After making this curious comparison they has-
tened sunward, the guide promising to join us

shortly on the bank of a deep pool, where we were to wait for him. This is a charming little lakelet of unknown depth, never yet stirred by a breeze, and its eternal calm excites the imagination even more profoundly than the silvery lakes of the glaciers rimmed with meadows and snow and reflecting sublime mountains.

Our guide, a jolly, rollicking Italian, led us into the heart of the hill, up and down, right and left, from chamber to chamber more and more magnificent, all a-glitter like a glacier cave with icicle-like stalactites and stalagmites combined in forms of indescribable beauty. We were shown one large room that was occasionally used as a dancing-hall; another that was used as a chapel, with natural pulpit and crosses and pews, sermons in every stone, where a priest had said mass. Mass-saying is not so generally developed in connection with natural wonders as dancing. One of the first conceits excited by the giant Sequoias was to cut one of them down and dance on its stump. We have also seen dancing in the spray of Niagara; dancing in the famous Bower Cave above Coulterville; and nowhere have I seen so much dancing as in Yosemite. A dance on the inaccessible South Dome would likely follow the making of an easy way to the top of it.

It was delightful to witness here the infinite deliberation of Nature, and the simplicity of her methods in the production of such mighty results, such perfect repose combined with restless enthusiastic energy. Though cold and bloodless as a landscape of polar ice, building was going on in the dark with incessant activity. The archways and ceilings were everywhere hung with down-growing crystals, like inverted groves of leafless saplings, some of them large, others delicately attenuated, each tipped with a single drop of water, like the terminal bud of a pine-tree. The only appreciable sounds were the dripping and tinkling of water falling into pools or faintly plashing on the crystal floors.

In some places the crystal decorations are arranged in graceful flowing folds deeply plicated like stiff silken drapery. In others straight lines of the ordinary stalactite forms are combined with reference to size and tone in a regularly graduated system like the strings of a harp with musical tones corresponding thereto; and on these stone harps we played by striking the crystal strings with a stick. The delicious liquid tones they gave forth seemed perfectly divine as they sweetly whispered and wavered through the majestic halls and died away in faintest cadence,—the music of fairy-land. Here we lingered and reveled, rejoicing to find so much music in stony silence, so much splendor in darkness, so many mansions in the depths of the mountains, buildings ever in process of construction, yet ever finished, developing from perfection to perfection, profusion without overabundance; every particle visible or invisible in glorious motion, marching to the music of the spheres in a region regarded as the abode of eternal stillness and death.

The outer chambers of mountain caves are frequently selected as homes by wild beasts. In the

Sierra, however, they seem to prefer homes and hiding-places in chaparral and beneath shelving precipices, as I have never seen their tracks in any of the caves. This is the more remarkable because notwithstanding the darkness and oozing water there is nothing uncomfortably cellar-like or sepulchral about them.

When we emerged into the bright landscapes of the sun everything looked brighter, and we felt our faith in Nature's beauty strengthened, and saw more clearly that beauty is universal and immortal, above, beneath, on land and sea, mountain and plain, in heat and cold, light and darkness.

CHAPTER XVI

THE BEE-PASTURES

WHEN California was wild, it was one sweet bee-garden throughout its entire length, north and south, and all the way across from the snowy Sierra to the ocean.

Wherever a bee might fly within the bounds of this virgin wilderness—through the redwood forests, along the banks of the rivers, along the bluffs and headlands fronting the sea, over valley and plain, park and grove, and deep, leafy glen, or far up the piny slopes of the mountains—throughout every belt and section of climate up to the timber line, bee-flowers bloomed in lavish abundance. Here they grew more or less apart in special sheets and patches of no great size, there in broad, flowing folds hundreds of miles in length—zones of polleny forests, zones of flowery chaparral, stream-tangles of rubus and wild rose, sheets of golden compositæ, beds of violets, beds of mint, beds of bryanthus and clover, and so on, certain species blooming somewhere all the year round.

But of late years plows and sheep have made sad havoc in these glorious pastures, destroying tens of thousands of the flowery acres like a fire, and banishing many species of the best honey-plants to rocky cliffs and fence-corners, while, on the other hand, cultivation thus far has given no adequate compensation, at least in kind; only acres of alfalfa for miles of the richest wild pasture, ornamental roses and honeysuckles around cottage doors for cascades of wild roses in the dells, and small, square orchards and orange-groves for broad mountain-belts of chaparral.

The Great Central Plain of California, during the months of March, April, and May, was one smooth, continuous bed of honey-bloom, so marvelously rich that, in walking from one end of it to the other, a distance of more than 400 miles, your foot would press about a hundred flowers at every step. Mints, gilias, nemophilas, castilleias, and innumerable compositæ were so crowded together that, had ninety-nine per cent. of them been taken away, the plain would still have seemed to any but Californians extravagantly flowery. The radiant, honeyful corollas, touching and overlapping, and rising above one another, glowed in the living light like a sunset sky—one sheet of purple and gold, with the bright Sacramento pouring through the midst of it from the north, the San Joaquin from the south, and their many tributaries sweeping in at right angles from the mountains, dividing the plain into sections fringed with trees.

Along the rivers there is a strip of bottom-land, countersunk beneath the general level, and wider toward the foot-hills, where magnificent oaks, from three to eight feet in diameter, cast grateful masses of shade over the open, prairie-like levels. And close along the water's edge there was a fine jungle of tropical luxuriance, composed of wild-rose and bramble bushes and a great variety of climbing vines, wreathing and interlacing the branches and trunks of willows and alders, and swinging across from summit to summit in heavy festoons. Here the wild bees reveled in fresh bloom long after the flowers of the drier plain had withered and gone to seed. And in midsummer, when the " blackberries " were ripe, the Indians came from the mountains to feast—men, women, and babies in long, noisy trains, often joined by the farmers of the neighborhood, who gathered this wild fruit with commendable appreciation of its superior flavor, while their home orchards were full of ripe peaches, apricots, nectarines, and figs, and their vineyards were laden with grapes. But, though these luxuriant, shaggy river-beds were thus distinct from the smooth, treeless plain, they made no heavy dividing lines in general views. The whole appeared as one continuous sheet of bloom bounded only by the mountains.

When I first saw this central garden, the most extensive and regular of all the bee-pastures of the State, it seemed all one sheet of plant gold, hazy and vanishing in the distance, distinct as a new map along the foot-hills at my feet.

Descending the eastern slopes of the Coast Range through beds of gilias and lupines, and around many a breezy hillock and bush-crowned headland, I at length waded out into the midst of it. All the ground was covered, not with grass and green leaves, but with radiant corollas, about ankle-deep next the foot-hills, knee-deep or more five or six

A BEE-RANCH IN LOWER CALIFORNIA.

miles out. Here were bahia, madia, madaria, burrielia, chrysopsis, corethrogyne, grindelia, etc.,
growing in close social congregations of various
shades of yellow, blending finely with the purples
of clarkia, orthocarpus, and œnothera, whose delicate petals were drinking the vital sunbeams without giving back any sparkling glow.

Because so long a period of extreme drought
succeeds the rainy season, most of the vegetation
is composed of annuals, which spring up simultaneously, and bloom together at about the same height
above the ground, the general surface being but
slightly ruffled by the taller phacelias, pentstemons,
and groups of *Salvia carduacea,* the king of the mints.

Sauntering in any direction, hundreds of these
happy sun-plants brushed against my feet at every
step, and closed over them as if I were wading in
liquid gold. The air was sweet with fragrance, the
larks sang their blessed songs, rising on the wing as
I advanced, then sinking out of sight in the polleny sod, while myriads of wild bees stirred the
lower air with their monotonous hum — monotonous, yet forever fresh and sweet as every-day sunshine. Hares and spermophiles showed themselves
in considerable numbers in shallow places, and
small bands of antelopes were almost constantly in
sight, gazing curiously from some slight elevation,
and then bounding swiftly away with unrivaled
grace of motion. Yet I could discover no crushed
flowers to mark their track, nor, indeed, any destructive action of any wild foot or tooth whatever.

The great yellow days circled by uncounted,
while I drifted toward the north, observing the

countless forms of life thronging about me, lying down almost anywhere on the approach of night. And what glorious botanical beds I had! Oftentimes on awaking I would find several new species leaning over me and looking me full in the face, so that my studies would begin before rising.

About the first of May I turned eastward, crossing the San Joaquin River between the mouths of the Tuolumne and Merced, and by the time I had reached the Sierra foot-hills most of the vegetation had gone to seed and become as dry as hay.

All the seasons of the great plain are warm or temperate, and bee-flowers are never wholly wanting; but the grand springtime — the annual resurrection — is governed by the rains, which usually set in about the middle of November or the beginning of December. Then the seeds, that for six months have lain on the ground dry and fresh as if they had been gathered into barns, at once unfold their treasured life. The general brown and purple of the ground, and the dead vegetation of the preceding year, give place to the green of mosses and liverworts and myriads of young leaves. Then one species after another comes into flower, gradually overspreading the green with yellow and purple, which lasts until May.

The "rainy season" is by no means a gloomy, soggy period of constant cloudiness and rain. Perhaps nowhere else in North America, perhaps in the world, are the months of December, January, February, and March so full of bland, plant-building sunshine. Referring to my notes of the winter and spring of 1868–69, every day of which I spent out of doors, on that section of the plain lying between the Tuolumne and Merced rivers, I find that the first rain of the season fell on December 18th. January had only six rainy days — that is, days on which rain fell; February three, March five, April three, and May three, completing the so-called rainy season, which was about an average one. The ordinary rain-storm of this region is seldom very cold or violent. The winds, which in settled weather come from the northwest, veer round into the opposite direction, the sky fills gradually and evenly with one general cloud, from which the rain falls steadily, often for days in succession, at a temperature of about 45° or 50°.

More than seventy-five per cent. of all the rain of this season came from the northwest, down the coast over southeastern Alaska, British Columbia, Washington, and Oregon, though the local winds of these circular storms blow from the southeast. One magnificent local storm from the northwest fell on March 21. A massive, round-browed cloud came swelling and thundering over the flowery plain in most imposing majesty, its bossy front burning white and purple in the full blaze of the sun, while warm rain poured from its ample fountains like a cataract, beating down flowers and bees, and flooding the dry watercourses as suddenly as those of Nevada are flooded by the so-called "cloud-bursts." But in less than half an hour not a trace of the heavy, mountain-like cloud-structure was left in the sky, and the bees were on the wing, as if nothing more gratefully refreshing could have been sent them.

By the end of January four species of plants were in flower, and five or six mosses had already adjusted their hoods and were in the prime of life; but the flowers were not sufficiently numerous as yet to affect greatly the general green of the young leaves. Violets made their appearance in the first week of February, and toward the end of this month the warmer portions of the plain were already golden with myriads of the flowers of rayed compositæ.

This was the full springtime. The sunshine grew warmer and richer, new plants bloomed every day; the air became more tuneful with humming wings, and sweeter with the fragrance of the opening flowers. Ants and ground squirrels were getting ready for their summer work, rubbing their benumbed limbs, and sunning themselves on the husk-piles before their doors, and spiders were busy mending their old webs, or weaving new ones.

In March, the vegetation was more than doubled in depth and color; claytonia, calandrinia, a large white gilia, and two nemophilas were in bloom, together with a host of yellow compositæ, tall enough now to bend in the wind and show wavering ripples of shade.

In April, plant-life, as a whole, reached its greatest height, and the plain, over all its varied surface, was mantled with a close, furred plush of purple and golden corollas. By the end of this month, most of the species had ripened their seeds, but undecayed, still seemed to be in bloom from the numerous corolla-like involucres and whorls of chaffy scales of the compositæ. In May, the bees found in flower only a few deep-set liliaceous plants and eriogonums.

June, July, August, and September is the season of rest and sleep,— a winter of dry heat,— followed in October by a second outburst of bloom at the very driest time of the year. Then, after the shrunken mass of leaves and stalks of the dead vegetation crinkle and turn to dust beneath the foot, as if it had been baked in an oven, *Hemizonia virgata*, a slender, unobtrusive little plant, from six inches to three feet high, suddenly makes its appearance in patches miles in extent, like a resurrection of the bloom of April. I have counted upward of 3000 flowers, five eighths of an inch in diameter, on a single plant. Both its leaves and stems are so slender as to be nearly invisible, at a distance of a few yards, amid so showy a multitude of flowers. The ray and disk flowers are both yellow, the stamens purple, and the texture of the rays is rich and velvety, like the petals of garden pansies. The prevailing wind turns all the heads round to the southeast, so that in facing northwestward we have the flowers looking us in the face. In my estimation, this little plant, the last born of the brilliant host of compositæ that glorify the plain, is the most interesting of all. It remains in flower until November, uniting with two or three species of wiry eriogonums, which continue the floral chain around December to the spring flowers of January. Thus, although the main bloom and honey season is only about three months long, the floral circle, however thin around some of the hot, rainless months, is never completely broken.

How long the various species of wild bees have lived in this honey-garden, nobody knows; probably ever since the main body of the present flora gained possession of the land, toward the close of the glacial period. The first brown honey-bees brought to California are said to have arrived in San Francisco in March, 1853. A bee-keeper by the name of Shelton purchased a lot, consisting of twelve swarms, from some one at Aspinwall, who had brought them from New York. When landed at San Francisco, all the hives contained live bees, but they finally dwindled to one hive, which was taken to San José. The little immigrants flourished and multiplied in the bountiful pastures of the Santa Clara Valley, sending off three swarms the first season. The owner was killed shortly afterward, and in settling up his estate, two of the swarms were sold at auction for $105 and $110 respectively. Other importations were made, from time to time, by way of the Isthmus, and, though great pains were taken to insure success, about one half usually died on the way. Four swarms were brought safely across the plains in 1859, the hives being placed in the rear end of a wagon, which was stopped in the afternoon to allow the bees to fly and feed in the floweriest places that were within reach until dark, when the hives were closed.

In 1855, two years after the time of the first arrivals from New York, a single swarm was brought over from San José, and let fly in the Great Central Plain. Bee-culture, however, has never gained much attention here, notwithstanding the extraor-

dinary abundance of honey-bloom, and the high
price of honey during the early years. A few hives
are found here and there among settlers who
chanced to have learned something about the business before coming to the State. But sheep, cattle,
grain, and fruit raising are the chief industries, as
they require less skill and care, while the profits
thus far have been greater. In 1856 honey sold here
at from one and a half to two dollars per pound.
Twelve years later the price had fallen to twelve
and a half cents. In 1868 I sat down to dinner
with a band of ravenous sheep-shearers at a ranch
on the San Joaquin, where fifteen or twenty hives
were kept, and our host advised us not to spare the
large pan of honey he had placed on the table, as
it was the cheapest article he had to offer. In all
my walks, however, I have never come upon a regular bee-ranch in the Central Valley like those so
common and so skilfully managed in the southern
counties of the State. The few pounds of honey
and wax produced are consumed at home, and are
scarcely taken into account among the coarser
products of the farm. The swarms that escape from
their careless owners have a weary, perplexing
time of it in seeking suitable homes. Most of
them make their way to the foot-hills of the mountains, or to the trees that line the banks of the
rivers, where some hollow log or trunk may be
found. A friend of mine, while out hunting on
the San Joaquin, came upon an old coon trap,
hidden among some tall grass, near the edge of the
river, upon which he sat down to rest. Shortly
afterward his attention was attracted to a crowd

of angry bees that were flying excitedly about his head, when he discovered that he was sitting upon their hive, which was found to contain more than 200 pounds of honey. Out in the broad, swampy delta of the Sacramento and San Joaquin rivers, the little wanderers have been known to build their combs in a bunch of rushes, or stiff, wiry grass, only slightly protected from the weather, and in danger every spring of being carried away by floods. They have the advantage, however, of a vast extent of fresh pasture, accessible only to themselves.

The present condition of the Grand Central Garden is very different from that we have sketched. About twenty years ago, when the gold placers had been pretty thoroughly exhausted, the attention of fortune-seekers — not home-seekers — was, in great part, turned away from the mines to the fertile plains, and many began experiments in a kind of restless, wild agriculture. A load of lumber would be hauled to some spot on the free wilderness, where water could be easily found, and a rude box-cabin built. Then a gang-plow was procured, and a dozen mustang ponies, worth ten or fifteen dollars apiece, and with these hundreds of acres were stirred as easily as if the land had been under cultivation for years, tough, perennial roots being almost wholly absent. Thus a ranch was established, and from these bare wooden huts, as centers of desolation, the wild flora vanished in ever-widening circles. But the arch destroyers are the shepherds, with their flocks of hoofed locusts, sweeping over the ground like a fire, and trampling down every rod that escapes the plow as completely as if the whole plain were a cottage garden-plot without a fence.   But notwithstanding these destroyers, a thousand swarms of bees may be pastured here for every one now gathering honey. The greater portion is still covered every season with a repressed growth of bee-flowers, for most of the species are annuals, and many of them are not relished by sheep or cattle, while the rapidity of their growth enables them to develop and mature their seeds before any foot has time to crush them.   The ground is, therefore, kept sweet, and the race is perpetuated, though only as a suggestive shadow of the magnificence of its wildness.

The time will undoubtedly come when the entire area of this noble valley will be tilled like a garden, when the fertilizing waters of the mountains, now flowing to the sea, will be distributed to every acre, giving rise to prosperous towns, wealth, arts, etc. Then, I suppose, there will be few left, even among botanists, to deplore the vanished primeval flora. In the mean time, the pure waste going on—the wanton destruction of the innocents—is a sad sight to see, and the sun may well be pitied in being compelled to look on.

The bee-pastures of the Coast Ranges last longer and are more varied than those of the great plain, on account of differences of soil and climate, moisture, and shade, etc.   Some of the mountains are upward of 4000 feet in height, and small streams, springs, oozy bogs, etc., occur in great abundance and variety in the wooded regions, while open parks, flooded with sunshine, and hill-girt valleys lying at different elevations, each with its own peculiar climate and exposure, possess the required conditions for the development of species and families of plants widely varied.

Next the plain there is, first, a series of smooth hills, planted with a rich and showy vegetation that differs but little from that of the plain itself — as if the edge of the plain had been lifted and bent into flowing folds, with all its flowers in place, only toned down a little as to their luxuriance, and a few new species introduced, such as the hill lupines, mints, and gilias. The colors show finely when thus held to view on the slopes; patches of red, purple, blue, yellow, and white, blending around the edges, the whole appearing at a little distance like a map colored in sections.

Above this lies the park and chaparral region, with oaks, mostly evergreen, planted wide apart, and blooming shrubs from three to ten feet high; manzanita and ceanothus of several species, mixed with rhamnus, cercis, pickeringia, cherry, amelanchier, and adenostoma, in shaggy, interlocking thickets, and many species of hosackia, clover, monardella, castilleia, etc., in the openings.

The main ranges send out spurs somewhat parallel to their axes, inclosing level valleys, many of them quite extensive, and containing a great profusion of sun-loving bee-flowers in their wild state; but these are, in great part, already lost to the bees by cultivation.

Nearer the coast are the giant forests of the redwoods, extending from near the Oregon line to Santa Cruz. Beneath the cool, deep shade of these majestic trees the ground is occupied by ferns, chiefly woodwardia and aspidiums, with only a few flowering plants—oxalis, trientalis, erythronium, fritillaria, smilax, and other shade-lovers. But all along the redwood belt there are sunny openings on hill-slopes looking to the south, where the giant trees stand back, and give the ground to the small sunflowers and the bees. Around the lofty redwood walls of these little bee-acres there is usually a fringe of Chestnut Oak, Laurel, and Madroño, the last of which is a surpassingly beautiful tree, and a great favorite with the bees. The trunks of the largest specimens are seven or eight feet thick, and about fifty feet high; the bark red and chocolate colored, the leaves plain, large, and glossy, like those of *Magnolia grandiflora*, while the flowers are yellowish-white, and urn-shaped, in well-proportioned panicles, from five to ten inches long. When in full bloom, a single tree seems to be visited at times by a whole hive of bees at once, and the deep hum of such a multitude makes the listener guess that more than the ordinary work of honey-winning must be going on.

How perfectly enchanting and care-obliterating are these withdrawn gardens of the woods—long vistas opening to the sea—sunshine sifting and pouring upon the flowery ground in a tremulous, shifting mosaic, as the light-ways in the leafy wall open and close with the swaying breeze—shining leaves and flowers, birds and bees, mingling together in springtime harmony, and soothing fragrance exhaling from a thousand thousand fountains! In these balmy, dissolving days, when the deep heart-beats of Nature are felt thrilling rocks and trees and everything alike, common business and friends are happily forgotten, and even the natural honey-work of bees, and the care of birds for their young, and mothers for their children, seem slightly out of place.

To the northward, in Humboldt and the adjacent counties, whole hillsides are covered with rhododendron, making a glorious melody of bee-bloom in the spring. And the Western azalea, hardly less flowery, grows in massy thickets three to eight feet high around the edges of groves and woods as far south as San Luis Obispo, usually accompanied by manzanita; while the valleys, with their varying moisture and shade, yield a rich variety of the smaller honey-flowers, such as mentha, lycopus, micromeria, audibertia, trichostema, and other mints; with vaccinium, wild strawberry, geranium, calais, and goldenrod; and in the cool glens along the stream-banks, where the shade of trees is not too deep, spiræa, dog-wood, heteromeles, and calycanthus, and many species of rubus form interlacing tangles, some portion of which continues in bloom for months.

Though the coast region was the first to be invaded and settled by white men, it has suffered less from a bee point of view than either of the other main divisions, chiefly, no doubt, because of the unevenness of the surface, and because it is owned and protected instead of lying exposed to the flocks of the wandering "sheepmen." These remarks apply more particularly to the north half of the coast. Farther south there is less moisture, less forest shade, and the honey flora is less varied.

23

The Sierra region is the largest of the three main divisions of the bee-lands of the State, and the most regularly varied in its subdivisions, owing to their gradual rise from the level of the Central Plain to the alpine summits. The foot-hill region is about as dry and sunful, from the end of May until the setting in of the winter rains, as the plain. There are no shady forests, no damp glens, at all like those lying at the same elevations in the Coast Mountains. The social compositæ of the plain, with a few added species, form the bulk of the herbaceous portion of the vegetation up to a height of 1500 feet or more, shaded lightly here and there with oaks and Sabine Pines, and interrupted by patches of ceanothus and buckeye. Above this, and just below the forest region, there is a dark, heath-like belt of chaparral, composed almost exclusively of *Adenostoma fasciculata,* a bush belonging to the rose family, from five to eight feet high, with small, round leaves in fascicles, and bearing a multitude of small white flowers in panicles on the ends of the upper branches. Where it occurs at all, it usually covers all the ground with a close, impenetrable growth, scarcely broken for miles.

Up through the forest region, to a height of about 9000 feet above sea-level, there are ragged patches of manzanita, and five or six species of ceanothus, called deer-brush or California lilac. These are the most important of all the honey-bearing bushes of the Sierra. *Chamæbatia foliolosa,* a little shrub about a foot high, with flowers like the strawberry, makes handsome carpets beneath the pines, and seems to be a favorite with the bees; while pines themselves furnish unlimited quantities of pollen and honey-dew. The product of a single tree, ripening its pollen at the right time of year, would be sufficient for the wants of a whole hive. Along the streams there is a rich growth of lilies, larkspurs, pedicularis, castilleias, and clover. The alpine region contains the flowery glacier meadows, and countless small gardens in all sorts of places full of potentilla of several species, spraguea, ivesia, epilobium, and goldenrod, with beds of bryanthus and the charming cassiope covered with sweet bells. Even the tops of the mountains are blessed with flowers,—dwarf phlox, polemonium, ribes, hulsea, etc. I have seen wild bees and butterflies feeding at a height of 13,000 feet above the sea. Many, however, that go up these dangerous heights never come down again. Some, undoubtedly, perish in storms, and I have found thousands lying dead or benumbed on the surface of the glaciers, to which they had perhaps been attracted by the white glare, taking them for beds of bloom.

From swarms that escaped their owners in the lowlands, the honey-bee is now generally distributed throughout the whole length of the Sierra, up to an elevation of 8000 feet above sea-level. At this height they flourish without care, though the snow every winter is deep. Even higher than this several bee-trees have been cut which contained over 200 pounds of honey.

The destructive action of sheep has not been so general on the mountain pastures as on those of the great plain, but in many places it has been

more complete, owing to the more friable character
of the soil, and its sloping position. The slant
digging and down-raking action of hoofs on the
steeper slopes of moraines has uprooted and buried many of the tender plants from year to
year, without allowing them time to mature their
seeds. The shrubs, too, are badly bitten, especially
the various species of ceanothus. Fortunately,
neither sheep nor cattle care to feed on the manzanita, spiræa, or adenostoma; and these fine honey-
bushes are too stiff and tall, or grow in places too
rough and inaccessible, to be trodden under foot.
Also the cañon walls and gorges, which form so
considerable a part of the area of the range, while
inaccessible to domestic sheep, are well fringed
with honey-shrubs, and contain thousands of
lovely bee-gardens, lying hid in narrow side-cañons
and recesses fenced with avalanche taluses, and on
the top of flat, projecting headlands, where only
bees would think to look for them.

But, on the other hand, a great portion of the
woody plants that escape the feet and teeth of the
sheep are destroyed by the shepherds by means of
running fires, which are set everywhere during the
dry autumn for the purpose of burning off the old
fallen trunks and underbrush, with a view to im-
proving the pastures, and making more open ways
for the flocks. These destructive sheep-fires sweep
through nearly the entire forest belt of the range,
from one extremity to the other, consuming not
only the underbrush, but the young trees and seedlings on which the permanence of the forests depends; thus setting in motion a long train of evils

WILD BEE GARDEN

which will certainly reach far beyond bees and bee-keepers.

The plow has not yet invaded the forest region to any appreciable extent, neither has it accomplished much in the foot-hills. Thousands of bee-ranches might be established along the margin of the plain, and up to a height of 4000 feet, wherever water could be obtained. The climate at this elevation admits of the making of permanent homes, and by moving the hives to higher pastures as the lower pass out of bloom, the annual yield of honey would be nearly doubled. The foot-hill pastures, as we have seen, fail about the end of May, those of the chaparral belt and lower forests are in full bloom in June, those of the upper and alpine region in July, August, and September. In Scotland, after the best of the Lowland bloom is past, the bees are carried in carts to the Highlands, and set free on the heather hills. In France, too, and in Poland, they are carried from pasture to pasture among orchards and fields in the same way, and along the rivers in barges to collect the honey of the delightful vegetation of the banks. In Egypt they are taken far up the Nile, and floated slowly home again, gathering the honey-harvest of the various fields on the way, timing their movements in accord with the seasons. Were similar methods pursued in California the productive season would last nearly all the year.

The average elevation of the north half of the Sierra is, as we have seen, considerably less than that of the south half, and small streams, with the bank and meadow gardens dependent upon them,

are less abundant. Around the head waters of the
Yuba, Feather, and Pitt rivers, the extensive table-
lands of lava are sparsely planted with pines,
through which the sunshine reaches the ground
with little interruption. Here flourishes a scattered, tufted growth of golden applopappus, linosyris, bahia, wyetheia, arnica, artemisia, and similar
plants; with manzanita, cherry, plum, and thorn
in ragged patches on the cooler hill-slopes. At the
extremities of the Great Central Plain, the Sierra
and Coast Ranges curve around and lock together
in a labyrinth of mountains and valleys, throughout which their floras are mingled, making at the
north, with its temperate climate and copious rainfall, a perfect paradise for bees, though, strange to
say, scarcely a single regular bee-ranch has yet been
established in it.

Of all the upper flower fields of the Sierra,
Shasta is the most honeyful, and may yet surpass
in fame the celebrated honey hills of Hybla and
hearthy Hymettus. Regarding this noble mountain from a bee point of view, encircled by its
many climates, and sweeping aloft from the torrid plain into the frosty azure, we find the first
5000 feet from the summit generally snow-clad,
and therefore about as honeyless as the sea. The
base of this arctic region is girdled by a belt of
crumbling lava measuring about 1000 feet in vertical breadth, and is mostly free from snow in
summer. Beautiful lichens enliven the faces of the
cliffs with their bright colors, and in some of the
warmer nooks there are a few tufts of alpine daisies,
wall-flowers and pentstemons; but, notwithstanding

these bloom freely in the late summer, the zone as a whole is almost as honeyless as the icy summit, and its lower edge may be taken as the honey-line. Immediately below this comes the forest zone, covered with a rich growth of conifers, chiefly Silver Firs, rich in pollen and honey-dew, and diversified with countless garden openings, many of them less than a hundred yards across. Next, in orderly succession, comes the great bee zone. Its area far surpasses that of the icy summit and both the other zones combined, for it goes sweeping majestically around the entire mountain, with a breadth of six or seven miles and a circumference of nearly a hundred miles.

Shasta, as we have already seen, is a fire-mountain created by a succession of eruptions of ashes and molten lava, which, flowing over the lips of its several craters, grew outward and upward like the trunk of a knotty exogenous tree. Then followed a strange contrast. The glacial winter came on, loading the cooling mountain with ice, which flowed slowly outward in every direction, radiating from the summit in the form of one vast conical glacier — a down-crawling mantle of ice upon a fountain of smoldering fire, crushing and grinding for centuries its brown, flinty lavas with incessant activity, and thus degrading and remodeling the entire mountain. When, at length, the glacial period began to draw near its close, the ice-mantle was gradually melted off around the bottom, and, in receding and breaking into its present fragmentary condition, irregular rings and heaps of moraine matter were stored upon its flanks. The glacial erosion of most of the Shasta lavas produces detritus, composed of rough, sub-angular boulders of moderate size and of porous gravel and sand, which yields freely to the transporting power of running water. Magnificent floods from the ample fountains of ice and snow working with sublime energy upon this prepared glacial detritus, sorted it out and carried down immense quantities from the higher slopes, and reformed it in smooth, delta-like beds around the base; and it is these flood-beds joined together that now form the main honey-zone of the old volcano.

Thus, by forces seemingly antagonistic and destructive, has Mother Nature accomplished her beneficent designs—now a flood of fire, now a flood of ice, now a flood of water; and at length an outburst of organic life, a milky way of snowy petals and wings, girdling the rugged mountain like a cloud, as if the vivifying sunbeams beating against its sides had broken into a foam of plant-bloom and bees, as sea-waves break and bloom on a rock shore.

In this flowery wilderness the bees rove and revel, rejoicing in the bounty of the sun, clambering eagerly through bramble and hucklebloom, ringing the myriad bells of the manzanita, now humming aloft among polleny willows and firs, now down on the ashy ground among gilias and buttercups, and anon plunging deep into snowy banks of cherry and buckthorn. They consider the lilies and roll into them, and, like lilies, they toil not, for they are impelled by sun-power, as water-wheels by water-power; and when the one has plenty of high-pressure water, the other plenty of sunshine, they hum and quiver alike. Sauntering in the Shasta bee-lands in the sun-days of summer, one may readily infer the time of day from the comparative energy of bee-movements alone—drowsy and moderate in the cool of the morning, increasing in energy with the ascending sun, and, at high noon, thrilling and quivering in wild ecstasy, then gradually declining again to the stillness of night. In my excursions among the glaciers I occasionally meet bees that are hungry, like mountaineers who venture too far and remain too long above the bread-line; then they droop and wither like autumn leaves. The Shasta bees are perhaps better fed than any others in the Sierra. Their field-work is one perpetual feast; but, however exhilarating the sunshine or bountiful the supply of flowers, they are always dainty feeders. Humming-moths and humming-birds seldom set foot upon a flower, but poise on the wing in front of it, and reach forward as if they were sucking through straws. But bees, though as dainty as they, hug their favorite flowers with profound cordiality, and push their blunt, polleny faces against them, like babies on their mother's bosom. And fondly, too, with eternal love, does Mother Nature clasp her small bee-babies, and suckle them, multitudes at once, on her warm Shasta breast.

Besides the common honey-bee there are many other species here—fine mossy, burly fellows, who were nourished on the mountains thousands of sunny seasons before the advent of the domestic species. Among these are the bumblebees, mason-bees, carpenter-bees, and leaf-cutters. Butterflies, too, and moths of every size and pattern; some broad-winged like bats, flapping slowly, and sailing in easy curves; others like small, flying violets, shaking about loosely in short, crooked flights close to the flowers, feasting luxuriously night and day. Great numbers of deer also delight to dwell in the brushy portions of the bee-pastures.

Bears, too, roam the sweet wilderness, their blunt, shaggy forms harmonizing well with the trees and tangled bushes, and with the bees, also, notwithstanding the disparity in size. They are fond of all good things, and enjoy them to the utmost, with but little troublesome discrimination — flowers and leaves as well as berries, and the bees themselves as well as their honey. Though the California bears have as yet had but little experience with honey-bees, they often succeed in reaching their bountiful stores, and it seems doubtful whether bees themselves enjoy honey with so great a relish. By means of their powerful teeth and claws they can gnaw and tear open almost any hive conveniently accessible. Most honey-bees, however, in search of a home are wise enough to make choice of a hollow in a living tree, a considerable distance above the ground, when such places are to be had; then they are pretty secure, for though the smaller black and brown bears climb well, they are unable to break into strong hives while compelled to exert themselves to keep from falling, and at the same time to endure the stings of the fighting bees without having their paws free to rub them off. But woe to the black bumblebees discovered in their mossy nests in the ground! With a few strokes of their huge paws the bears uncover the entire establishment, and, before time is given for a general buzz, bees old and young, larvæ, honey, stings, nest, and all are taken in one ravishing mouthful.

Not the least influential of the agents concerned in the superior sweetness of the Shasta flora are its storms—storms I mean that are strictly local, bred and born on the mountain. The magical rapidity with which they are grown on the mountain-top, and bestow their charity in rain and snow, never fails to astonish the inexperienced lowlander. Often in calm, glowing days, while the bees are still on the wing, a storm-cloud may be seen far above in the pure ether, swelling its pearl bosses, and growing silently, like a plant. Presently a clear, ringing discharge of thunder is heard, followed by a rush of wind that comes sounding over the bending woods like the roar of the ocean, mingling rain-drops, snow-flowers, honey-flowers, and bees in wild storm harmony.

Still more impressive are the warm, reviving days of spring in the mountain pastures. The blood of the plants throbbing beneath the life-giving sun-shine seems to be heard and felt. Plant growth goes on before our eyes, and every tree in the woods, and every bush and flower is seen as a hive of restless industry. The deeps of the sky are mottled with singing wings of every tone and color; clouds of brilliant chrysididæ dancing and swirling in exquisite rhythm, golden-barred vespidæ, dragon-flies, butterflies, grating cicadas, and jolly, rattling grasshoppers, fairly enameling the light.

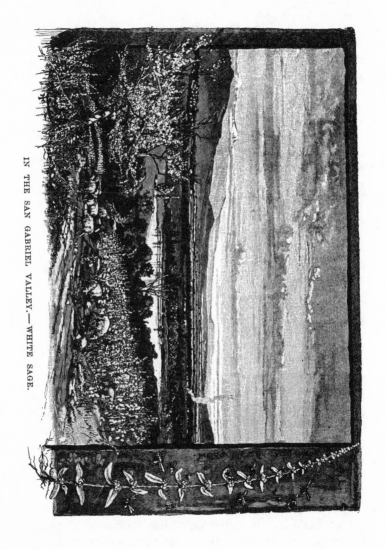

IN THE SAN GABRIEL VALLEY.—WHITE SAGE.

On bright, crisp mornings a striking optical effect may frequently be observed from the shadows of the higher mountains while the sunbeams are pouring past overhead. Then every insect, no matter what may be its own proper color, burns white in the light. Gauzy-winged hymenoptera, moths, jet-black beetles, all are transfigured alike in pure, spiritual white, like snowflakes.

In Southern California, where bee-culture has had so much skilful attention of late years, the pasturage is not more abundant, or more advantageously varied as to the number of its honey-plants and their distribution over mountain and plain, than that of many other portions of the State where the industrial currents flow in other channels. The famous White Sage (*Audibertia*), belonging to the mint family, flourishes here in all its glory, blooming in May, and yielding great quantities of clear, pale honey, which is greatly prized in every market it has yet reached. This species grows chiefly in the valleys and low hills. The Black Sage on the mountains is part of a dense, thorny chaparral, which is composed chiefly of adenostoma, ceanothus, manzanita, and cherry—not differing greatly from that of the southern portion of the Sierra, but more dense and continuous, and taller, and remaining longer in bloom. Streamside gardens, so charming a feature of both the Sierra and Coast Mountains, are less numerous in Southern California, but they are exceedingly rich in honey-flowers, wherever found,—melilotus, columbine, collinsia, verbena, zauschneria, wild rose, honeysuckle, philadelphus, and lilies rising from the warm, moist dells in a very storm of exuberance. Wild buckwheat of many species is developed in abundance over the dry, sandy valleys and lower slopes of the mountains, toward the end of summer, and is, at this time, the main dependence of the bees, reinforced here and there by orange groves, alfalfa fields, and small home gardens.

The main honey months, in ordinary seasons, are April, May, June, July, and August; while the other months are usually flowery enough to yield sufficient for the bees.

According to Mr. J. T. Gordon, President of the Los Angeles County Bee-keepers' Association, the first bees introduced into the county were a single hive, which cost $150 in San Francisco, and arrived in September, 1854.[1] In April, of the following year, this hive sent out two swarms, which were sold for $100 each. From this small beginning the bees gradually multiplied to about 3000 swarms in the year 1873. In 1876 it was estimated that there were between 15,000 and 20,000 hives in the county, producing an annual yield of about 100 pounds to the hive—in some exceptional cases, a much greater yield.

In San Diego County, at the beginning of the season of 1878, there were about 24,000 hives, and the shipments from the one port of San Diego for the same year, from July 17 to November 10, were 1071 barrels, 15,544 cases, and nearly 90 tons. The

---

[1] Fifteen hives of Italian bees were introduced into Los Angeles County in 1855, and in 1876 they had increased to 500. The marked superiority claimed for them over the common species is now attracting considerable attention.

largest bee-ranches have about a thousand hives, and are carefully and skilfully managed, every scientific appliance of merit being brought into use. There are few bee-keepers, however, who own half as many as this, or who give their undivided attention to the business. Orange culture, at present, is heavily overshadowing every other business.

A good many of the so-called bee-ranches of Los Angeles and San Diego counties are still of the rudest pioneer kind imaginable. A man unsuccessful in everything else hears the interesting story of the profits and comforts of bee-keeping, and concludes to try it; he buys a few colonies, or gets them from some overstocked ranch on shares, takes them back to the foot of some cañon, where the pasturage is fresh, squats on the land, with, or without, the permission of the owner, sets up his hives, makes a box-cabin for himself, scarcely bigger than a bee-hive, and awaits his fortune.

Bees suffer sadly from famine during the dry years which occasionally occur in the southern and middle portions of the State. If the rainfall amounts only to three or four inches, instead of from twelve to twenty, as in ordinary seasons, then sheep and cattle die in thousands, and so do these small, winged cattle, unless they are carefully fed, or removed to other pastures. The year 1877 will long be remembered as exceptionally rainless and distressing. Scarcely a flower bloomed on the dry valleys away from the stream-sides, and not a single grain-field depending upon rain was reaped. The seed only sprouted, came up a little way, and

A BEE-RANCH ON A SPUR OF THE SAN GABRIEL RANGE.
CARDINAL FLOWER.

24

withered. Horses, cattle, and sheep grew thinner
day by day, nibbling at bushes and weeds, along
the shallowing edges of streams, many of which
were dried up altogether, for the first time since
the settlement of the country.

In the course of a trip I made during the summer of that year through Monterey, San Luis
Obispo, Santa Barbara, Ventura, and Los Angeles
counties, the deplorable effects of the drought were
everywhere visible — leafless fields, dead and dying
cattle, dead bees, and half-dead people with dusty,
doleful faces. Even the birds and squirrels were
in distress, though their suffering was less pain-
fully apparent than that of the poor cattle. These
were falling one by one in slow, sure starvation
along the banks of the hot, sluggish streams, while
thousands of buzzards correspondingly fat were
sailing above them, or standing gorged on the
ground beneath the trees, waiting with easy faith
for fresh carcasses. The quails, prudently considering the hard times, abandoned all thought of
pairing. They were too poor to marry, and so
continued in flocks all through the year without
attempting to rear young. The ground-squirrels,
though an exceptionally industrious and enterprising race, as every farmer knows, were hard pushed
for a living; not a fresh leaf or seed was to be
found save in the trees, whose bossy masses of
dark green foliage presented a striking contrast
to the ashen baldness of the ground beneath them.
The squirrels, leaving their accustomed feeding-
grounds, betook themselves to the leafy oaks to
gnaw out the acorn stores of the provident wood-

WILD BUCKWHEAT.—A BEE-RANCH IN THE WILDERNESS.

peckers, but the latter kept up a vigilant watch
upon their movements. I noticed four woodpeckers
in league against one squirrel, driving the poor
fellow out of an oak that they claimed. He dodged
round the knotty trunk from side to side, as
nimbly as he could in his famished condition, only
to find a sharp bill everywhere. But the fate of
the bees that year seemed the saddest of all.
In different portions of Los Angeles and San Diego
counties, from one half to three fourths of them
died of sheer starvation. Not less than 18,000
colonies perished in these two counties alone, while
in the adjacent counties the death-rate was hardly
less.

Even the colonies nearest to the mountains suffered this year, for the smaller vegetation on the
foot-hills was affected by the drought almost as
severely as that of the valleys and plains, and even
the hardy, deep-rooted chaparral, the surest dependence of the bees, bloomed sparingly, while
much of it was beyond reach. Every swarm could
have been saved, however, by promptly supplying
them with food when their own stores began to fail,
and before they became enfeebled and discouraged;
or by cutting roads back into the mountains, and
taking them into the heart of the flowery chaparral. The Santa Lucia, San Rafael, San Gabriel,
San Jacinto, and San Bernardino ranges are almost
untouched as yet save by the wild bees. Some
idea of their resources, and of the advantages and
disadvantages they offer to bee-keepers, may be
formed from an excursion that I made into the
San Gabriel Range about the beginning of August

of "the dry year." This range, containing most of
the characteristic features of the other ranges just
mentioned, overlooks the Los Angeles vineyards
and orange groves from the north, and is more
rigidly inaccessible in the ordinary meaning of the
word than any other that I ever attempted to
penetrate. The slopes are exceptionally steep
and insecure to the foot, and they are covered with
thorny bushes from five to ten feet high. With
the exception of little spots not visible in general
views, the entire surface is covered with them,
massed in close hedge growth, sweeping gracefully
down into every gorge and hollow, and swelling
over every ridge and summit in shaggy, ungovernable exuberance, offering more honey to the acre
for half the year than the most crowded clover-
field. But when beheld from the open San Gabriel
Valley, beaten with dry sunshine, all that was seen
of the range seemed to wear a forbidding aspect.
From base to summit all seemed gray, barren,
silent, its glorious chaparral appearing like dry
moss creeping over its dull, wrinkled ridges and
hollows.

Setting out from Pasadena, I reached the foot of
the range about sundown; and being weary and
heated with my walk across the shadeless valley,
concluded to camp for the night. After resting a
few moments, I began to look about among the
flood-boulders of Eaton Creek for a camp-ground,
when I came upon a strange, dark-looking man
who had been chopping cord-wood. He seemed
surprised at seeing me, so I sat down with him on
the live-oak log he had been cutting, and made

haste to give a reason for my appearance in his solitude, explaining that I was anxious to find out something about the mountains, and meant to make my way up Eaton Creek next morning. Then he kindly invited me to camp with him, and led me to his little cabin, situated at the foot of the mountains, where a small spring oozes out of a bank overgrown with wild-rose bushes. After supper, when the daylight was gone, he explained that he was out of candles; so we sat in the dark, while he gave me a sketch of his life in a mixture of Spanish and English. He was born in Mexico, his father Irish, his mother Spanish. He had been a miner, rancher, prospector, hunter, etc., rambling always, and wearing his life away in mere waste; but now he was going to settle down. His past life, he said, was of "no account," but the future was promising. He was going to "make money and marry a Spanish woman." People mine here for water as for gold. He had been running a tunnel into a spur of the mountain back of his cabin. "My prospect is good," he said, "and if I chance to strike a good, strong flow, I 'll soon be worth $5000 or $10,000. For that flat out there," referring to a small, irregular patch of bouldery detritus, two or three acres in size, that had been deposited by Eaton Creek during some flood season,—"that flat is large enough for a nice orange-grove, and the bank behind the cabin will do for a vineyard, and after watering my own trees and vines I will have some water left to sell to my neighbors below me, down the valley. And then," he continued, "I can keep bees, and make money

A BEE-PASTURE ON THE MORAINE DESERT. SPANISH BAYONET.

that way, too, for the mountains above here are
just full of honey in the summer-time, and one of
my neighbors down here says that he will let me
have a whole lot of hives, on shares, to start with.
You see I 've a good thing; I 'm all right now."
All this prospective affluence in the sunken,
boulder-choked flood-bed of a mountain-stream!
Leaving the bees out of the count, most fortune-
seekers would as soon think of settling on the
summit of Mount Shasta. Next morning, wishing
my hopeful entertainer good luck, I set out on my
shaggy excursion.

About half an hour's walk above the cabin, I
came to "The Fall," famous throughout the valley
settlements as the finest yet discovered in the San
Gabriel Mountains. It is a charming little thing,
with a low, sweet voice, singing like a bird, as it
pours from a notch in a short ledge, some thirty-
five or forty feet into a round mirror-pool. The
face of the cliff back of it, and on both sides, is
smoothly covered and embossed with mosses,
against which the white water shines out in showy
relief, like a silver instrument in a velvet case.
Hither come the San Gabriel lads and lassies, to
gather ferns and dabble away their hot holidays in
the cool water, glad to escape from their common-
place palm-gardens and orange-groves. The delicate
maidenhair grows on fissured rocks within reach
of the spray, while broad-leaved maples and syca-
mores cast soft, mellow shade over a rich profusion
of bee-flowers, growing among boulders in front of
the pool—the fall, the flowers, the bees, the ferny
rocks, and leafy shade forming a charming little

poem of wildness, the last of a series extending down the flowery slopes of Mount San Antonio through the rugged, foam-beaten bosses of the main Eaton Cañon.

From the base of the fall I followed the ridge that forms the western rim of the Eaton basin to the summit of one of the principal peaks, which is about 5000 feet above sea-level. Then, turning eastward, I crossed the middle of the basin, forcing a way over its many subordinate ridges and across its eastern rim, having to contend almost everywhere with the floweriest and most impenetrable growth of honey-bushes I had ever encountered since first my mountaineering began. Most of the Shasta chaparral is leafy nearly to the ground; here the main stems are naked for three or four feet, and interspiked with dead twigs, forming a stiff *chevaux de frise* through which even the bears make their way with difficulty. I was compelled to creep for miles on all fours, and in following the bear-trails often found tufts of hair on the bushes where they had forced themselves through.

For 100 feet or so above the fall the ascent was made possible only by tough cushions of club-moss that clung to the rock. Above this the ridge weathers away to a thin knife-blade for a few hundred yards, and thence to the summit of the range it carries a bristly mane of chaparral. Here and there small openings occur on rocky places, commanding fine views across the cultivated valley to the ocean. These I found by the tracks were favorite outlooks and resting-places for the wild animals—bears, wolves, foxes, wildcats, etc.—which abound here, and would have to be taken into account in the establishment of bee-ranches. In the deepest thickets I found wood-rat villages—groups of huts four to six feet high, built of sticks and leaves in rough, tapering piles, like musk-rat cabins. I noticed a good many bees, too, most of them wild. The tame honey-bees seemed languid and wing-weary, as if they had come all the way up from the flowerless valley.

After reaching the summit I had time to make only a hasty survey of the basin, now glowing in the sunset gold, before hastening down into one of the tributary cañons in search of water. Emerging from a particularly tedious breadth of chaparral, I found myself free and erect in a beautiful park-like grove of Mountain Live Oak, where the ground was planted with aspidiums and brier-roses, while the glossy foliage made a close canopy overhead, leaving the gray dividing trunks bare to show the beauty of their interlacing arches. The bottom of the cañon was dry where I first reached it, but a bunch of scarlet mimulus indicated water at no great distance, and I soon discovered about a bucketful in a hollow of the rock. This, however, was full of dead bees, wasps, beetles, and leaves, well steeped and simmered, and would, therefore, require boiling and filtering through fresh charcoal before it could be made available. Tracing the dry channel about a mile farther down to its junction with a larger tributary cañon, I at length discovered a lot of boulder pools, clear as crystal, brimming full, and linked together by glistening streamlets just strong enough to sing audibly. Flowers in

A BEE-KEEPER'S CABIN.—BURRIELIA (ABOVE).—MADIA (BELOW).

full bloom adorned their margins, lilies ten feet high, larkspur, columbines, and luxuriant ferns, leaning and overarching in lavish abundance, while a noble old Live Oak spread its rugged arms over all. Here I camped, making my bed on smooth cobblestones.

Next day, in the channel of a tributary that heads on Mount San Antonio, I passed about fifteen or twenty gardens like the one in which I slept—lilies in every one of them, in the full pomp of bloom. My third camp was made near the middle of the general basin, at the head of a long system of cascades from ten to 200 feet high, one following the other in close succession down a rocky, inaccessible cañon, making a total descent of nearly 1700 feet. Above the cascades the main stream passes through a series of open, sunny levels, the largest of which are about an acre in size, where the wild bees and their companions were feasting on a showy growth of zauschneria, painted cups, and monardella; and gray squirrels were busy harvesting the burs of the Douglas Spruce, the only conifer I met in the basin.

The eastern slopes of the basin are in every way similar to those we have described, and the same may be said of other portions of the range. From the highest summit, far as the eye could reach, the landscape was one vast bee-pasture, a rolling wilderness of honey-bloom, scarcely broken by bits of forest or the rocky outcrops of hilltops and ridges.

Behind the San Bernardino Range lies the wild "sage-brush country," bounded on the east by the Colorado River, and extending in a general northerly direction to Nevada and along the eastern base of the Sierra beyond Mono Lake.